T0190234

The Elements of Knowledge Organization

Richard P. Smiraglia

The Elements of Knowledge Organization

Richard P. Smiraglia
University of Wisconsin-Milwaukee
Milwaukee, WI, USA

ISBN 978-3-319-36264-9   ISBN 978-3-319-09357-4 (eBook)
DOI 10.1007/978-3-319-09357-4
Springer Cham Heidelberg New York Dordrecht London

© Springer International Publishing Switzerland 2014
Softcover reprint of the hardcover 1st edition 2014
This work is subject to copyright. All rights are reserved by the Publisher, whether the whole or part of the material is concerned, specifically the rights of translation, reprinting, reuse of illustrations, recitation, broadcasting, reproduction on microfilms or in any other physical way, and transmission or information storage and retrieval, electronic adaptation, computer software, or by similar or dissimilar methodology now known or hereafter developed. Exempted from this legal reservation are brief excerpts in connection with reviews or scholarly analysis or material supplied specifically for the purpose of being entered and executed on a computer system, for exclusive use by the purchaser of the work. Duplication of this publication or parts thereof is permitted only under the provisions of the Copyright Law of the Publisher's location, in its current version, and permission for use must always be obtained from Springer. Permissions for use may be obtained through RightsLink at the Copyright Clearance Center. Violations are liable to prosecution under the respective Copyright Law.
The use of general descriptive names, registered names, trademarks, service marks, etc. in this publication does not imply, even in the absence of a specific statement, that such names are exempt from the relevant protective laws and regulations and therefore free for general use.
While the advice and information in this book are believed to be true and accurate at the date of publication, neither the authors nor the editors nor the publisher can accept any legal responsibility for any errors or omissions that may be made. The publisher makes no warranty, express or implied, with respect to the material contained herein.

Printed on acid-free paper

Springer is part of Springer Science+Business Media (www.springer.com)

Contents

Chapter 1
Introduction: An Overview of Knowledge Organization

The photos above are views of one restored corner of one part of the Minoan palace at Knossos on Crete. Here is another view of the corner of the palace:

This is a great example of what it is like to work in knowledge organization. Sometimes we see one entity, sometimes if we are fortunate we can see that same entity from different points of view. Sometimes, if we can step back a little bit, we can understand the entity by seeing it adjacent to other entities in the same domain. Look again at the last photo—to the upper right you see the hillside into which the palace was built. Had I wanted, I could have shown the Aegean Sea by including a view down the hill and to the left, behind the corner, as it were. We see only a tiny bit at a time of anything, and we understand even less. The search for meaning is critical. But it must always be like this—stepwise, a little to the left, a little to the right, look up from below, look all around, and so on.

1.1   The Beginning: Science and Technology in Relation

If you have been introduced to survey courses in information or knowledge organization then you have some acquaintance with the tools for organizing knowledge for information retrieval (subject headings, classifications, catalogs, thesauri, taxonomies, ontologies, etc.). These two concepts, then, make sort of an expression:

Organizing knowledge <—> Information retrieval

Knowledge, or that which is known, can be organized in various ways. Some of the ways in which we organize knowledge involve heuristics, or natural rules, and some of the ways we use are pragmatic. Once the knowledge has been ordered in some way, then it is available to be retrieved by susceptible users, for whom it can become information (Buckland 1988). These techniques constitute the apparatus of knowledge organization, much of which is at the core of librarianship. Librarianship is one technology that is based on the science of information.

But the science of knowledge itself can be approached in a variety of ways. I will approach it from the point of view of information, and that will color our vision somewhat. Every discipline has its own approach to the science of knowledge—sometimes called typology, sometimes called taxonomy, sometimes called ontology—and that very interesting distinction will help us understand our own role as purveyors of the substrate (Bates 1999) of information. Science at the most basic level is simply the act of research, which itself is the act of self-conscious inquiry. The results of scientific inquiry become the content of the discipline itself. So the discipline now known as knowledge organization is the sum of the research discovered about the conceptual ordering of knowledge and about the bridge across disciplines that allows us to view the effective substrate.

Point of view is critical. If we seek to comprehend knowledge ontologically, it means we seek to do so from a universal point of view in which we can identify and position all entities relative to one another. If we seek to comprehend knowledge through typology, it means we are empirically identifying entities as we discover them and grouping them according to characteristics as best we can. If we seek to comprehend knowledge taxonomically, it means we are working with meaning, seeking to understand the whole by first finding and defining its parts. None of these points of view have yet been related to information retrieval; rather, all of them illustrate the ways in which knowledge organization is critical for all scholarly endeavor. If, then, we seek to comprehend knowledge as an entity for information retrieval, it means we are working with repositories of documents containing recorded knowledge; our job will be to extract precisely relevant bits of that which is known for later use. In all aspects the point of view is critical. It is not such a simple thing just to make a list of subject headings, or just to classify books by placing them in broad categories (as libraries do). Nor is it a simple thing to classify disease or race or even groceries (Bowker and Star 1999).

Knowledge organization is critical for the proper functioning of the science of information. Without that which is learned in KO, information retrieval cannot work. But the science of knowledge organization is clearly the province of different philosophical points of view. Which means that, in the end, information retrieval is only as efficacious as the understanding of KO. The technology is, therefore, critically subject to the science on which it rests.

1.2   Therefore Knowledge Is?

We must begin by asking these questions:

1.2.1   How Do I Know?

The question "how do I know?" forms the basis of epistemology, the science of knowing, which itself forms one of the central tenets of knowledge organization. Before we can understand how knowledge is intrinsically ordered we must first understand the point of view from which knowledge is perceived.

1.2.2   What Is?

By asking "what is?" we turn to the other cornerstone of knowledge organization, ontology, or the science of being. Because ordering requires some degree of categorization, which is a form of determining likeness, we must create rules for what "is" or "is not" included. Inclusion implies exclusion, and these are the first elements in any ordering of knowledge.

1.2.3   How Is It Ordered?

Knowledge structures for conceptual ordering become critical once knowledge itself is perceived. At a meta-level both recorded and unrecorded knowledge can be defined empirically. Taxonomy is a framework in which elements are defined, and categories are mutually exclusive and collectively exhaustive; typology is a parallel framework in which elements are categorized functionally and share characteristics empirically. At the domain level knowledge used by a discourse community can be framed by a structured ontology, or represented symbolically with a classification, or rendered in the form of a controlled vocabulary, such as a thesaurus. At the artifactual level, individual bits of recorded knowledge are controlled using knowledge representation schema such as metadata. These are the elements of knowledge organization systems.

1.3   About This book

This book is organized according to the outline presented above. First I discuss concepts of "theory" at a metalevel, and then I look the historical path that has led to the evolution of knowledge organization, first as a documentary practice, and more recently as a science itself. Next I look specifically at the core elements of knowledge organization, epistemology and ontology. Finally I look closely at the specific elements of knowledge organization: metadata, taxonomy, classification, domain analysis, and thesauri. All of it is dependent on point of view.

References

Bates, Marcia J. 1999. The invisible substrate of information science. Journal of the American Society for Information Science 50: 1043–50.

Bowker, Geoffrey C., and Susan Leigh Star. 1999. Sorting things out: classification and its consequences. Cambridge: MIT Press.

Buckland, Michael K. 1988. Library services in theory and context, 2nd ed. Oxford: Pergamon Press.

Chapter 2
About Theory of Knowledge Organization

2.1 On Theory

To understand the importance of theory in the development of scientific thought one has to rely on a thorough comprehension of the tools and paradigms of research. At the most basic level, theory is a frequently-tested (and thereby affirmed) statement of the interacting requirements of a phenomenon. In empirical research, theory is both the accumulated wisdom of the paradigm from which hypotheses are cast and the constant reaccumulation that occurs as each hypothesis is tested. The essence of empirical theory is the notion that probability theory allows us to state with great precision the degree to which our statements likely mirror reality. In other domains theories have more the aura of accumulated statements that describe positions within a system. In sum, the presence of a theoretical basis in a domain, whether a single theory or a system of theoretical statements, implies not just the cleverness of the actors in the domain, but rather their scientific productivity. Theory exists in domains where a large quantity of research has been very productive at generating workable explanations and also at identifying inadequate or erroneous statements.

So if there were to be a theory of knowledge organization what would it look like? Obviously it would have to include operational definitions of both of the key terms—knowledge, and organization. It would have to supply environmental parameters within which the two phenomena interact. And it would have to describe the manner in which these phenomena interact. In essence, a theory of knowledge organization would have to explain the impact of the organization of knowledge on those for whom it is operationalized, whether animate or not.

There are, in fact, several theoretical contributions that seek to explain knowledge organization. I will review four discrete points of view in this essay, in order to

Portions of this text appeared as Chapter 1. Introduction: theory, knowledge organization, epistemology, culture. In Smiraglia, Richard P. and Hur-Li Lee eds., 2012. *Cultural frames of knowledge*. Würzburg: Ergon-Verlag, pp. 1–17. Reprinted by permission.

© Springer International Publishing Switzerland 2014
R.P. Smiraglia, *The Elements of Knowledge Organization*,
DOI 10.1007/978-3-319-09357-4_2

arrive at an overview that will help us consider how current research is contributing to theory in the domain. Dahlberg (2006) was the founder of the domain as we now know it. In particular she was the founder of the International Society for Knowledge Organization. Her point of view lays the groundwork for a particular approach to empirical analysis using the term "concept theoretic." I will begin with her ideas, because in many ways they are the most concise.

But we must look also at three different and all influential points of view. Patrick Wilson posed the backdrop of a bibliographical universe of texts in which various approaches to ordering might be found. He gave us a theoretical yardstick for evaluating the efficacy of all approaches—he called this exploitative power. If it is working it is powerfully driving the evolution of new knowledge, and that has important social consequences. More recently Elaine Svenonius attempted an explanation of the totality of organization of knowledge, by using a linguistic metaphor and designating a set-theoretic. Falling chronologically between the two we find Birger Hjørland's application of activity theory as an explanation for the phenomena of knowledge organization. I will first review the major thrust of these three texts, and then look at two articles of my own in which I attempted a summary of empirical evidence, and two articles by Hjørland that helped move Dahlberg's theoretic closer to fruition.

2.2   Dahlberg

It takes some bravery to put your ideas before the critical eyes of peers, and it often ends with difficulty. Peers see what they want to see, and often miss critical points, no matter how carefully crafted. So it is a testimony to Dahlberg that she sought to turn the mostly rationalist/pragmatist act of classification into the science of the order of knowledge. There are several earlier papers that established her goals, but in 2006 she offered the paper cited here for publication to help explain some of the most basic (and most misunderstood) tenets of knowledge organization. Let us consider it her epistle to the post-modern ISKO domain. In this paper she answered all of our theoretical questions. To wit (Dahlberg 2006, 12):

knowledge = the known
organization = the activity of constructing something according to a plan.

To elucidate what she means by knowledge, she explains further that knowledge may be transferred in space and time, and is dependent on language. Note that this is an utterly social definition, which restricts knowledge to the human dimension. In this theory, knowledge is a commodity of humans that is shared with purpose, and therefore is not raw, nor is it unattached to a human thought, nor is it unutterable. For Dahlberg, knowledge exists only in the dimension of human perception. She says there are four ways in which it can be perceived:

– Knowledge elements (characteristics of concepts);
– Knowledge units (concepts);

- Larger knowledge units (concept combinations); and,
- Knowledge systems (knowledge units arranged in a planned, cohesive structure).

For example, the temperature is high, flames are leaping about, matter is being consumed—these are elements of the knowledge of fire, which is a concept. Pistons work, fuel is consumed, wheels turn, firemen ride—these are characteristics of the engine of a fire department. We may combine these knowledge units, or concepts— of fire, and engine—into a concept combination (or a term) "fire engine." Furthermore, we can create a small hierarchy with two classes and a rule of synthesis, such that:

1: Fire: high temperature, flames, consumption of matter
2: Engine: pistons, fuel, firemen, wheels, ride

Add any n to any other n in natural linguistic sequence if a sensical result ensues

1-2: Fire engine

In this manner we have created a knowledge organization system (the ubiquitous KOS), by the use of deliberate planning, and cohesive structure.

For Dahlberg, this process is the essence of knowledge organization. The process is constrained by human experience and bounded by linguistic borders. The process is semiotically dynamic, and can be repeated infinitely until everything is contained in one or more systems and all systems are linked. In fact, to ground the process, Dahlberg also identifies three approaches to the designation of concepts (Dahlberg 2006, 13):

Mathematical-statistical: cluster analysis of terms;
Mathematical-conceptual: lattice theory for visual graphing of relationships;
Concept-theoretical: analyses the contents of concepts.

Notice that the latter approach is not explained. We can imagine use of co-word analysis (mathematical statistical) or of multi-dimensional scaling (mathematical-conceptual), and in fact, bibliometric methods use these techniques to generate taxonomies that describe the axes of domains. But, the final approach, which is the key to Dahlberg's science, is the most elusive. We will hold this thought while we turn back to Wilson's bibliographical universe. It will be Hjørland's appeal to activity theory that will flesh out an operational plan for concept-theoretic.

2.3   Wilson

Two Kinds of Power is an immensely influential book (see Smiraglia 2007), that has fueled more than a generation of research in knowledge organization, and in information retrieval. In it, Wilson elucidated the dichotomous goals of controlling recorded knowledge as over against the creation of new knowledge. His theoretical construct was presented in the form of a philosophical bibliographical essay. The citations are far-ranging and the footnotes entertainingly expressive. He captured the frustration of scholars attempting to interact with the known universe of fact, even as they themselves create new, frustratingly complex, material. The power of this theoretical explanation is its universality and its presentation in natural language. But make no mistake, his terms are operational and have been used for decades to generate research (see for example Mai 2011, 2013; Smiraglia and van den Heuvel 2013).

2.3.1 The Bibliographical Universe

The central part of Wilson's theory is his conception of the bibliographical universe as a concept space wherein one might find in orbit or transit all exemplars of recorded knowledge. Wilson at once sets his sights only on recorded knowledge—this sets his notion apart from some aspects of Dahlberg's, because nothing is included that has not been recorded (recorded texts, therefore, can be retrieved). To wit (Wilson 1968, 6): "The totality of things over which bibliographical control is or might be exercised, consists of writings and recorded sayings." Of course, the physical universe is full of knowledge that is recorded in DNA and molecular structures and other sources, but these are not necessarily accessible to humans, being literate merely in their own tongues. Wilson frees the bibliographical apparatus from the linear existence it had up to this point. Instead of a vast index or card file, Wilson sees points in this universe orbiting and clustering and crossing the bibliographical macrocosm, in concert with each other according to specifiable (if so far unspecified) relationship patterns. Just as the physical universe reels with gravity and physical forces that propel, impel, and compel planets, stars, asteroids and other bodies to exist in relation to each other, so Wilson sees the bibliographical universe as a multi- dimensional, relational system. His mystical explanation goes no farther, but was inspiring enough to lead decades of scholars to seek explanations that might further describe his universe.

In Wilson's universe there are two domains or concept spaces (he calls them powers or controls; we might also think of them as dimensions)—which he calls descriptive and exploitative. The descriptive domain is the dimension where people labor to make indexes and catalogs of all of the texts of knowledge that they know to be extant. The exploitative domain is where scholars toil to create new knowledge by synthesizing that which already is known. It is very difficult to explain this differential. To librarians or archivists (especially catalogers) it seems he is referring to the cataloging department on the one hand and the users on the other. But he really means it in quite a different way. The descriptive domain is that place where what is known and already has been synthesized is described—so this includes not just indexes and catalogs, but also encyclopedias, textbooks, databases, the memories of scholars, and everything that in some way records that which already is known and synthesized. This is no simple list of raw documents. Rather it is the entirety of what is known, in the form in which it has been filtered by scholars and cultures through the ages and passed to us to curate. And, the exploitative domain is not just a place where users pose queries. Rather, it is that place where, in order to arrive at the best solution, the scholar must find bits of knowledge that are related in a fundamental way but that are so disjoint that they might never appear to be similar at all.

Every scholar has these moments, and often refers to them as serendipity. These are the moments when, after toiling over a text for months, one goes to the farmer's market, and the color of the apples suddenly reminds one of something that reminds one of something else that reminds one to go ask another question, and the answer to that question leads in a new direction where—bingo, one finds an amazing connection that now brings together two heretofore unrelated senses. That is what the exploitative domain is all about. Wilson is trying to say that catalogs and indexes are all very nice, and so are encyclopedias (and even mentor's memories), but, what scholars really need is some way to make the process less haphazard. If the bibliographic universe has bodies spinning in concert according to bibliographical laws, then let us describe all of those entities—the bodies and the laws—sufficiently that we might be able to predict relationships with accuracy.

The key to Wilson's theory is the concept of efficacy. Anything descriptive that makes exploitation possible is efficacious. That which is not efficacious is creating bibliographical drag on the system and should be expunged. This philosophical yardstick has been operationalized in many ways by researchers over the past four decades in order to justify the evolution of the bibliographical apparatus that we have today.

Oh yes, the bibliographical apparatus. Well, I have already described that as the product of the descriptive domain. Except, Wilson points out, the apparatus has rather the character of a *deus ex machina* (my interpretation, not Wilson's, by the way), which is to say, it is like a great big machine with certain cogs working perfectly and others rusted shut. One way of repairing the apparatus, according to Wilson, is by tending to the specifications of the various bibliographical instruments, and it is here that he attends to the pitfalls and joys of specific tools—indexes, bibliographies, catalogs, abstracts, and so forth. Notice that (p. 55): "Any text that refers in any way to any other text or copy of some text might be considered a potential bibliographical instrument." Even a simple citation, then, is a bibliographical instrument, much like a road sign.

Finally, Wilson excels in pointing out the linguistic disadvantages of conceptual systems. Subject analysis is fraught with phenomenological peril, and its product leads to various habits of hunting in order to couple appropriate references. It is not a pretty picture, as he points out the futility of a system built on assumptions about relevance, which (he says) does not really exist. He devotes an entire (the penultimate) chapter to the concept of reliability, foreshadowing another major work (Wilson 1983), *Second-hand knowledge: an inquiry into cognitive authority*. It is here, in his discussion of reliability, that he fleshes out the extension of what I have called efficacy (my word, not Wilson's). It is here that he points out the fact that no matter how elegant the apparatus, the true test is exploitative power, and there are few ways to measure such a thing with reliability. He says (p. 131):

> An estimate of power is an estimate of what one could do if one tried, of what success would be achieved in different attempts. The existence of multitudes of cases in which success cannot be recognized with certainty, or in which the very notion of success is of doubtful applicability, added to the obvious difficulties of estimating a power on the basis of a sample set of trials, effectively prevent such estimates, in the bibliographical case, from claiming exactitude or finality.

In the end, with what today seems a surprising bit of futuristic imagination, Wilson petitions a revelatory "Supreme Bibliographical Council," which will be able to decide which things known by what scholars when, might actually be related to each other and to a contemporary scholar's query. He suggests, and then rejects, the creation of a bibliographical policy that would collocate all results (a la Otlet's universal bibliographic control), in favor of a bibliographical policy for the rationalization (p. 144) of work of all sorts. If the test of a theoretical construct were simply its power to explain, the number of citations to Wilson's work (Smiraglia 2007) would be sufficient testimony. But the true test of a theoretical construct is its power to inspire—thus see the papers by Buckland and Shaw (2008) or Mai (2011, 2013) or the nascent work by Zherebchevsky et al. (2008)—we see at the remove of forty years from the introduction of Wilson's ideas and the beginning of the third generation of scholars to make reference to it (led, in these two cases by Wilson's contemporary Buckland (see Bates 2004), and Smiraglia, a disciple from the 1980s (see for example Smiraglia 1985), the power of this notion of rationalizing what is known to create better efficacy for the generation of new and necessary knowledge.

In the decades immediately following the publication of *Two Kinds of Power* two distinct research streams developed inspired by Wilson's vision. Information scientists, such as Belkin, Saracevic, Van Rijsbergen, Swanson and Bookstein (Smiraglia 2007, 11) sought to find answers to the first of Wilson's bibliographical policies—how can we collocate all like results? Another research stream developed around the problems of controlling that which is known in order to generate a better bibliographical apparatus. This stream has at its forefront Svenonius, Hjørland, and White. White, together with his Drexel University colleague Kathryn McCain, created the complex of techniques for extensive bibliometric analysis of domains; we will look at their work when we turn to informetrics and domain analysis in a subsequent chapter. But both Svenonius and Hjørland taught generations of new scholars, and both generated their own, more pragmatic, theoretical constructs for knowledge organization. We will look at both, working chronologically.

2.4 Svenonius

Elaine Svenonius was one of the twentieth century's most respected researchers in knowledge organization. A graduate of the empiricist school at the University of Chicago, her research was always tightly controlled and therefore highly reliable scientifically. In 2000 *The Intellectual Foundations of Information Organization* was published, containing her meta-construct for theory of knowledge organization. That the title of her book uses the phrase "information organization" instead of the term we are using (knowledge organization) is a sign of the imprecision of definitions within the discipline of information science and the sub-disciplines (or domains) that work within it. This is not the place to discuss the merits of these terms. Suffice it to say that both terms certainly are used, and with the same meaning, which is the organization of that which is known in order that it might be the product of the process of information retrieval.

Svenonius' framework begins with an outline of her intellectual foundation (p. 1), which includes an ideology of purposes and principles, the formalization of processes, research design, and key problems in need of resolution (Svenonius 2000). This is followed at once with an extensive historical analysis, which provides a precise set of parameters for the extension of the concept space in which she intends to work. That is, this is not the entire history of knowledge organization but it is the history of the precedents that yield Svenonius' theoretical construct. The second chapter is an analysis of bibliographic objectives, in which she clearly focuses her effort on the record of written knowledge to be found in bibliographical entities. And these bibliographical entities are the subject of the third chapter.

2.4.1 Set Theoretic

The first major element of her theoretical construct is her set theoretic, which is introduced almost accidentally within the discussion of entity types. She writes that (p. 35):

> Individual documents can be collected into *sets*, which themselves are bibliographic entities. Sets represent equivalence clusterings of documents. The individual members of a given set are equivalent with respect to the attributes they have in common. Potentially any attribute or collection of attributes can be used as a specification for set formation.

In this manner she maps a group of bibliographic typologies (about which more in a subsequent chapter)—categories that overlap and therefore are not mutually exclusive. Membership in any one category implies only clustering on the basis of the stated equivalence measure. Thus it is theoretically possible to isolate the attributes of a given bibliographic condition (my word, not hers) such as "origin" or "subject" the better to define the intension of each set over against the intensions of the other sets. Just as one might want a dress that also is red (thus borrowing from two types: clothing and color) so one might want a French translation of *Bleak House* (thus borrowing from two of Svenonius' sets: edition and superwork). Here are the five most important sets, which (she says) are mandated explicitly by the collocating objective (p. 35):

The set of all documents sharing essentially the same information (work)
The set of all documents sharing the same information (edition)

The set of all documents descended from a common origin (superwork)
The set of all documents by a given author
The set of all documents on a given subject.

In the next several chapters, Svenonius uses this set theoretic to describe how to operationalize bibliographical terminology. So, where Wilson had posed difficult questions and described the fuzziness of terminology, Svenonius now tries to supply a means for separating the intermingled attributes of entities so that they might be explicitly described. Potentially, this is a major step forward for research in knowledge organization. Unfortunately she does not continue to use the theoretic beyond this point in her text. Instead she turns to a set of linguistic metaphors.

2.4.2 Bibliographical Languages

The other major component of Svenonius' theoretical construct is itself a collocating device. Remember that to collocate is not only to draw things together, but to do so in order to disambiguate. Thus she suggests considering the domain of knowledge organization as a set of vocabularies with overlapping semantics, each of which might be considered its own language. The set of languages is (p. 54):

Work language
Author language
Title language
Edition language
Subject language
Classification language
Index language
Document language
Production language
Carrier language
Location language

Notice that she divides all Gaul into two parts—works and documents. This acknowledges the essential distinction between inventory control (document language) and intellectual access (work language), and it makes all aspects of intellectual access subordinate to the concept of the work. It is a quintessentially bibliographical point of view about the order of things, that all queries must eventually lead to "a work." Languages then have vocabulary, syntax, semantics, pragmatic uses, and rules. It is under "rules" that we find a partial (but telling) list of bibliographical standards. Here Svenonius has collocated the practice of bibliographical control—Wilson's bibliographical apparatus—as a pragmatic consequence of a post-modern Babel. Oh but that we all might speak one language!

The rest of Svenonius' book contains in-depth explanations of the set of languages in the list above. She attempts to broach this metaphorical Tower of Babel by clarifying the contents and the consequences of the plethora of bibliographical languages that constitute the bibliographical apparatus. While this book ends essentially without a conclusion—her "Afterword" is essentially a research agenda—we still have a concrete step forward in the statement of theory for knowledge organization. Svenonius' conception of the concept space, like Wilson's, is exclusively bibliographical and therefore the province of that which is known, synthesized, *and* recorded. The space is considered pragmatically from two perspectives, which might be thought to parallel Wilson's describing and exploiting. Specifically, Svenonius tells us to limit describing to document inventory, much of which can be automatic, and to focus instead on exploiting by expanding our conception of works and their attributes. She gives us two tools—a set theoretic, and a linguistic metaphor—with which to tackle this giant problem.

2.5 Hjørland

Birger Hjørland is arguably the most-cited author of theoretical work in the field of knowledge organization. His name frequently is found near or alongside Svenonius' in visualizations of the domain. And, again arguably, Hjørland has contributed the most directly usable applications analyses for the advance of knowledge organization. That is, his pragmatic writing urges authors in the domain to step aside from the pragmatic and to consider other epistemological perspectives. In 1997 his theoretical construct took form in the book *Information seeking and subject representation: an activity-theoretical approach to information science*. Here we see an appeal to understand documents not by their content but rather by the uses to which they are (or might be) put. This is not a new idea, for decades bibliographers (see Krummel 1976) have appealed to the notion that the actual physical form of documents is dictated by the marketplace and therefore the intellectual content also is molded by such considerations. This is an important principle for bibliography because it tells us to look beyond title pages for the clues to significant identification of specific documents as artifacts.

Here the thrust is different. Hjørland attempts to give an overview of information science based on the principle that information seeking is the key problem, over and against document representation. Thus his theoretical construct takes place entirely in Wilson's exploitative domain, leaving the descriptive domain for another day (or another author). His major thrust is subject searching and its requisite impact on the structure of information retrieval systems. Information seeking is presented from the point of view of "behavioral ecology," and he makes distinctions between documents and non-documents, and between known-item and unknown-item retrieval. Where Wilson posed a universe of writings, and Svenonius focused on documents only, Hjørland broadens the scope of the discipline to entities that record knowledge but that are not documents *per se*. Activity theory is clearly presented as a motivating factor in the metaphorical search for mushrooms (see p. 12 ff.), which draws convincing parallels. If we really want mushrooms we should be looking for the place with the best selection of mushrooms and not just the first batch we find under a tree. So, therefore, should searchers be locating their work according to the activity that drives it, in the best locations for good results. The anti-Google, we might call this. Knowledge organization is explicitly addressed in chapter 3, in relation to subject analysis. And the chapter after that outlines his reliance on epistemology.

2.5.1   Some Fundamentals

In 2003 Hjørland laid out some explicit marching orders for the domain of knowledge organization. Of particular importance was the new extension of the domain that he offered by extending it beyond the purview even of information science (as it traditionally has been understood) to the impact of the social division of labor and of social institutions. Principle actors in the domain are identified as knowledge producers and knowledge users. It is their two sets of activity that generate the dimensions of this universe. He is interested not just in indexing or document retrieval, but now also in scientific communication, the social roles of information, the epistemological stance of knowledge providers, and the impact of social semiotics. Hjørland's bibliographical universe is much broader than any we have seen before, and therefore the methodological requirements for research are all the less adequate.

2.6   Smiraglia, Hjørland

Is there a theory of knowledge organization? Not yet. There is, however, quite a lot of progress. In two papers, Smiraglia (2002a, b) used the tools of meta-analysis to suggest areas where empirical research has reached the level of theory. These are:

Author productivity and the distribution of name headings
The phenomenon of instantiation; and,
External validity.

The first two categories make liberal use of Lotka's Law to show that after several decades of empirical research it now is possible to predict the distribution of bibliographic phenomena in KOS if we know the bibliographic-demographic parameters of a set of documents (such as a library collection). The third category relies on the same bodies of research, to demonstrate that the bibliographic-demographics tell us that most libraries are, in fact, not just in supposition, alike. Thus research carried out in one library catalog, so long as the bibliographic-demographics are explicitly reported, can be generalized to other collections. There is potential theoretical predictive power in these results. The dimensions of the bibliographical universe can be not only comprehended but also recorded for exploitation. And with the wide comprehension of instantiation we see real evidence of what Svenonius' called the "Work language" and its impact on information retrieval. The extension of Lotka's Law from its original narrow use as predictor of author productivity to a new capability for demonstrating the extension of the bibliographic domain is also a major theoretical leap forward.

Hjørland (2008) brings our discussion full circle by acknowledging both broad and narrow definitions of the term knowledge organization. The narrow meaning is document description, the broad meaning is the social division of mental labor, the actual structure of that which is known and how it is conveyed in society. Thus we have Wilson's two powers—describing and exploiting—now defined as the extension of two dimensions of the power and use of knowledge. The impact of Dahlberg's concept-theoretic is its *use* in different domains.

References

Bates, Marcia J. 2004. Information science at the University of California at Berkeley in the 1960s: a memoir of student days. *Library trends* 52no4: 683–701.

Buckland, Michael C., and Ryan Shaw. 2008. 4W vocabulary mapping across diverse reference genres. In Arsenault, Clément and Joseph Tennis eds., *Culture and identity in knowledge organization: Proceedings of the 10th International ISKO Conference, Montréal, 5–8 August 2008.* Advances in knowledge organization 11. Würzburg: Ergon Verlag, pp. 151–56.

Dahlberg, Ingetraut. 2006. Knowledge organization: a new science? *Knowledge organization* 33: 11–19.

Hjørland. Birger. 1997. Information seeking and subject representation: an activity- theoretical approach to information science. New directions in information management 34. Westport, Conn.: Greenwood Press.

Hjørland. Birger. 2003. Some fundamentals of knowledge organization. *Knowledge organization* 30: 87–111.

Hjørland, Birger. 2008. What is knowledge organization (ko)? *Knowledge organization* 35: 86–101.

Krummel, D.W. 1976. Musical functions and bibliographical forms. *The library* 5th ser., 31: 327ff.

Mai, Jens-Erik. 2011. The modernity of classification. *Journal of documentation* 67: 710–30.

Mai, Jens-Erik. 2013. The quality and qualities of information. *Journal of the American Society for Information Science and Technology* 64: 675–88.

Smiraglia, Richard P. 1985. Theoretical considerations in the bibliographic control of music materials in libraries. *Cataloging & classification quarterly* 5n3:1–16.

Smiraglia, Richard P. 2002a. Progress toward theory in knowledge organization. *Library trends* 50: 300–49.

Smiraglia, Richard P. 2002b. Further progress toward theory in knowledge organization. *Canadian journal of information and library science.* 26 n2/3: 30–49.

Smiraglia, Richard P. 2007. Two Kinds of Power: insight into the legacy of Patrick Wilson. In Information Sharing in a Fragmented World: Crossing Boundaries: Proceedings of the Canadian Association for Information Science annual conference May 12–15, 2007, ed. Kimiz Dalkir and Clément Arsenault. http://www.cais-acsi.ca/2007proceedings.htm.

Smiraglia, Richard P., and Charles van den Heuvel. 2013. Classifications and concepts: toward an elementary theory of knowledge interaction. *Journal of documentation* 69: 360–83.

Svenonius, Elaine. 2000. *The intellectual foundation of information organization.* Cambridge, Mass.: MIT Press.

Wilson, Patrick. 1983. *Second-hand knowledge: an inquiry into cognitive authority.* Westport, Conn.: Greenwood Press.

Wilson, Patrick. 1968. *Two kinds of power: an essay in bibliographical control.* Berkeley: Univ. of California Press.

Zherebchevsky, Sergey, Nicolette Ceo, Michiko Tanaka, David Jank, Richard Smiraglia, and Stephen Stead. 2008. Classifying information objects: an exploratory ontological excursion. Poster presented at the 10th International ISKO Conference, Montréal, 5–8 August 2008.

Chapter 3
Philosophy: Underpinnings of Knowledge Organization

3.1 Why Philosophy?

As we saw in the chapter just past, the province of knowledge organization is much broader than many suspect. It is not simply the matter of indexing documents for retrieval. As Hjørland (2008) points out so eloquently, knowledge organization is closely related to the theory of knowledge itself, in a primary way. If the essential phenomenon of our domain is knowledge, then obvious questions arise as to what is known, and how it is known. The fundamental question of knowledge organization brings us to an even more basic level as we seek always to ask "what is?" Therefore, it is essential that we have a proper grounding in ontology (the study of being) and epistemology (the study of knowing), and we are best served as a multidisciplinary science by turning to philosophy for answers unfiltered by the activities of scholars in other domains touching on our own. Here I begin with some basic definitions that help us to understand the nature of knowledge, and therefore, of how it can be organized.

But also, along the way, lie three more areas rife for exploration. The first is related to epistemology. How do we know what it is that we know? Part of the answer lies in understanding how we as humans filter knowledge as we encounter it. We will look at theories of semiotics (signs) and phenomenology (perception) to find two sets of related answers to this question. The second question is what is order? We will look to Foucault in this connection, as we seek to find a post-modern system for the order of things. Finally, we will see how some of the work of Wittgenstein in the early twentieth century contributes to understanding of both sets of questions.

Classification, historically, has been the scholar's means for organized observation. At the most basic level, science is the art of classifying observations according

Portions of this text appeared as Chapter 1. Introduction: theory, knowledge organization, epistemology, culture. In Smiraglia, Richard P. and Hur-Li Lee eds. 2012. *Cultural frames of knowledge*. Würzburg: Ergon Verlag, pp. 1–17. Reprinted by permission.

© Springer International Publishing Switzerland 2014
R.P. Smiraglia, *The Elements of Knowledge Organization*,
DOI 10.1007/978-3-319-09357-4_3

to their likenesses and differences, which is essentially the meaning of ontology. What is, and what is not, constitute the boundaries or inclusion–exclusion criteria for categories. When categories together constitute a set of mutually exclusive and collectively exhaustive units, then they have evolved to the level of classification. A tiger is neither a turtle nor a boat, and we know this definitively because science has created precise inclusion–exclusion criteria for all three phenomena. Science is, therefore, the philosophy of comprehending the empirical. And knowledge organization is the science of the orderings of recorded phenomena. Like the science of information, knowledge organization relies on tools from a variety of other disciplines and therefore can be said to be pan-disciplinary, because it crosses all disciplines and also it can be said to be interdisciplinary, because it combines tools from different domains. The goal, ultimately, and the unity, is the ordering of phenomena.

3.1.1   Epistemology

Epistemology is the division of philosophy that investigates the nature and origin of knowledge. In philosophy at large, epistemology is central because it embraces the theory of knowledge itself. The central problems for epistemology are the definition of knowledge, and the means of its acquisition. The philosophical process engages a discourse in which skeptical challenges to any definition must be rebuked and therein lies the dilemma, for how can we study that which we cannot even define? According to Grayling (2003, 37) there are historically two chief schools of epistemological thought: rationalism and empiricism, which arise from mathematics and logic and the natural sciences, respectively. Epistemology begins with the simple definition that knowledge is justified true belief, and then proceeds to define the terms and to challenge them. Justification and belief yield when confronted with skepticism, and much modern philosophy (from Descartes and Locke forward) is concerned with the explanation of the components of this argument.

Although philosophers have identified many approaches to epistemology, in knowledge organization we have come to rely on a framework set forth by Birger Hjørland. (Together with Jeppe Nicolaisen, Hjørland has constructed a web tool called a "lifeboat"—a sort of web-based crib sheet—for epistemology. You can find it here: http://www.db.dk/jni/lifeboat/home.htm.) However, beginning from a basic metaphysical stance, Hjørland (1998, 608) lists four basic epistemological stances (or positions):

* Empiricism: derived from observation, perception, and experience;
* Rationalism: derived from the employment of reason over sensory experience;
* Historicism: derived from cultural hermeneutics; and,
* Pragmatism: derived from the consideration of goals and their consequences.

That which we know from our own experience of it, and in particular that which is known through the positivist sciences, is what we call empirical. We have solid evidence for the empirical, and we can point to the evidence as a means of prediction.

On the other hand, that which we know from reasoning about it, and in particular that which is known through the naturalistic sciences, is based in rationalism. There is no evidence, per se, for the rational; rather there are explanatory statements that seem to be logical when taken together. Historicist epistemology interprets evidence through a cultural lens, relying in particular on past experience. Pragmatism is exactly what it sounds like, derived from assumptions about the best means to an end. Pragmatic solutions work in the moment but do not necessarily rely on empirical evidence, and therefore do not necessarily pass the test of time. Rational solutions also often ignore empirical evidence and thus frequently yield unworkable schemas. Smiraglia (2002b) demonstrated the rational construction of catalogs based on assumptions but not on bibliographical evidence, but he also demonstrated an empirically-based structure that would be an improvement for the representation of instantiation. Later writings (2000, 2001a, b, 2002b, 2004, 2005a, b, 2006, 2008) used the same epistemological schema to demonstrate the breadth of the concept of instantiation among information objects.

Epistemology is an essential tool of knowledge organization, and the many papers that fall within its embrace at each international ISKO conference demonstrate its usefulness. Early papers of significance include Poli (1996), who contrasted the tools of ontology and epistemology for knowledge organization suggesting that where ontology represents the "objective" side of reality, epistemology represents the "subjective" side that allows for the perception of knowledge and its subjective role, and Olson (1996), who demonstrated Dewey's epistemic stance in the topography of recorded knowledge. In 2008 a section of the proceedings on epistemological foundations contained eleven contributions; in 2010 Hjørland led a separate seminar on epistemology, and another thirteen papers were found under the heading for epistemological foundations. In 2012 an anthology on epistemology in knowledge organization was published by Smiraglia and Lee; it was followed shortly by another anthology by Ibekwe-SanJaun and Dousa (2013). What we see clearly is that epistemology leads us to research questions about the essential nature of knowledge.

3.2 Semiotics: The Science or Theory of Signs

There is a natural affinity between the domains of knowledge and language, because language is the primary means by which knowledge is communicated among humans. Anyone who has ever tried to learn a new language has experienced the difficulty inherent in subtle shifts of meaning between cultures. Part of the problem lies in perception and prior experience; we will consider these issues when we discuss phenomenology later in this chapter. But meaning is itself a major component of the problem. The theories concerned with signs are attempts to describe some of the issues that underlie differences in meaning. We will look first at Ferdinand de Saussure's well-known notion about systems of signs. Then we will look at Charles

Sanders Peirce's theory of semiotics. These two conceptual representations are those that have had increasing numbers of adherents in the domain of knowledge organization.

There are, of course, other semiologists. Friedman (2008) includes a thorough overview of semiotic points of view that have been synthesized in knowledge organization. Umberto Eco is probably the author, aside from Peirce and Saussure, whose work finds frequent referents in knowledge organization. Eco's work arises from literary theory and embraces the concept of "open fields." Consider yourself standing at the edge of an open field. You might look across it to the other side where there are trees and a stream, or to the left where the railroad passes by. Or you might look into the field to see what is planted there. Or you might look down at the granularity of the surface, which itself is littered with manifold distinct phenomena. Eco says a text is like an open field and our experience of it, therefore, is personal, dynamic, and psychologically engaged. Morrisey (2002) has used Eco's "connotative semiotics" (Eco 1976) to analyze scientific works as multi-layered repositories of meaning that stretch from quantitative data points to declarative theories.

According to Malmkjær (2004, 465) linguistics can be seen as a subdivision of semiotics—the opposite of the point of view presented here—because semiotics is the study of signs, and linguistics is therefore concerned with the nature of linguistic signs. The process of making and using signs is semiosis; the term semiotic originated with Peirce; semiology is Saussure's term for the life of the sign in society (Malmkjær 2004, 466). Eco (1984, 4–7) referred to semiotics as specific or general, depending on whether the discourse was related to a particular system of signs or the whole study of the meaning of signs.

An interesting historical footnote concerns the chronology of these discoveries. Both Saussure and Peirce worked in the late nineteenth century, and in both cases their work was forgotten for nearly a century. It was not until the late twentieth century that scholars in other disciplines turned to semiotics to help understand meaning. One might hazard a guess that the rise of the Internet led to new necessity for understanding semantics. But likely there is more to it as well. It is also likely that scholarship needed to reach its moment of post-modern decomposition before scholars in diverse domains (such as musicology, and information, for instance) were forced to turn to semiotics for explanations. Nattiez (1990) and Goehr (1992) in musicology, for example, and Thomas and Smiraglia (1998) and Smiraglia (2002a) in information science, all used semiotic theory to discuss the nature of musical works as arbitrary auditory signifiers.

3.2.1 Saussure's Semiology

Ferdinand de Saussure was a Swiss scholar who is widely credited as the father of modern linguistics. His famous book *Course in general linguistics* (1959) was compiled from his lecture notes by former students after his death. The central concept of Saussure's linguistic theory was the concept of semiology, which is a system of signs that functions within society. Saussure's linguistics has a generic concern with texts and their interpretation, which makes his theory particularly amenable in the field of information, descended as it is from the field of documentation. He writes that language is a system of signs that express ideas, and therefore (p. 16) is "comparable to a system of writing." He says that linguistics is essential for understanding texts, which are the primary means by which knowledge advances in society (p. 7). Therefore, for Saussure, there is an intimate relationship between language, speech, and society, and this is best observable in texts. Saussure derives the name for this theory from the Greek "semeion," which means "sign," and suggests a theory of semiology could embrace the laws that govern signs as a consequence of social psychology (p. 16).

In Saussure's semiology the theory of signs is dyadic, meaning it has two components, which are the signifier and the signified. The sign itself is the unity of the two components. The signified is a concept, and the signifier is an associated sound-image. Saussure says that the signifier is immutable but the signified, which unfolds in time, is ultimately mutable. It is important to Saussure that the psychological aspects of the sign be considered, because he says (p. 65): "both terms involved in the linguistic sign are ... united in the brain by an associative bond." The "sound-image" is not a physical noise, but rather is the "psychological imprint" (p. 66), the impression made on the senses. For example: "the psychological character of our sound-images becomes apparent when we observe our own speech. Without moving our lips or tongue, we can talk to ourselves or recite mentally a selection of verse."

Saussure's sign has two principles, which he refers to as "primordial characteristics." The first is what he calls the arbitrary nature of the sign, by which he means the psychological association between the signifier and the signified. In other words, any sound-image may be associated with any concept. The fact that a particular sound-image becomes commonly associated with its attendant concept is an arbitrary consequence in time. Which leads to the second principle, the linear nature of the signifier, which unfolds in measurable time, and therefore is mutable because of the influence of the society in which it operates. Continuity in time, he says (p. 76) is coupled to change in time. Consider, for example, the word "gay." Two generations ago the word meant, as it had for more than a century, simply the concept of lively happiness. In the present generation the term is the preferred term for homosexual persons. The sound is the same, the signifier has changed. Language changes in time precisely because it becomes the property of the people who speak it.

3.2.2 Peirce's Semiotic

Semiotic theory originated with American philosopher Charles Sanders Peirce, who was a logician and mathematician by training, but who had an unfortunately checkered academic career. Because of his difficult professional life, much of his writing is either unpublished, or consists of unsynthesized notes gathered in volumes by the editors of his papers. Several of Peirce's discoveries are of major importance today,

Fig. 3.1 Peirce's sign

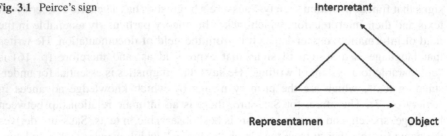

including not only semiotics, but also the philosophy of pragmatism, and the concept of electrical switching circuits that led to the development of digital technology in the mid-twentieth century. A review of Peirce's work reveals the interconnectedness of his thinking, which is a crucial point for understanding his philosophical positions. That is to say, all of his work is in essence a unity and therefore no single component can be taken in isolation without reference to other parts of his work. Semiotic theory is not simply the description of a sign as a concept dangling loosely in space. Rather, it is the description of the dynamic process of being in relation of any sort.

For Peirce, the sign consists of three components. These are the *Representamen*, the *Interpretant*, and the *Object*. The representamen is the concept as signal, the interpretant is the concept as reception, and the object is the concept as perception. Thus a sign is a process, which has famously been denoted thus (Fig. 3.1):

The key to the dynamism of Peirce's semiotic theory is the mutability of the object, which upon perception, becomes itself a new *representamen*. That which I say to you becomes your intellectual property once you comprehend it fully, and when you then express it, the process must necessarily begin again. Furthermore, Peirce says there are three kinds of signs, all of which are necessary to keep this dynamic process in motion. There are icons for likeness, signs that are similar or analogous to that which they represent, indexes, which are indicative signs that are somehow demonstrative of the phenomenon they represent (like a pronoun, Peirce says (1991, 181), which "forces the attention to the particular object intended without describing it)," and general signs, which are simply the names of symbols.

Peirce says (1991, 141–3) that a sign is "an object which stands for another to some mind." In order to qualify as a sign, there must be a real connection with the entity signified, so that the presence of the sign is clearly demonstrable. Furthermore, it must be regarded in a cognitive way as a sign, otherwise it will not function as a sign to human minds. See for example, Fig. 3.2, a photograph of a parking sign beside a houseboat on a canal in Amsterdam.

The parking sign is clearly a Peircian sign, because it has a clear connection to that which it represents and because it is recognizable as a sign. On the other hand, the geraniums on the roof of the houseboat, which often are literarily or metaphorically referred to as signs (as in, for example, "a sign of grace") are not a Peircian sign, because they do not have a literal, recognizable relationship to that for which they purportedly are signs. Elsewhere Peirce (1998, 4–5) says this is an important

Fig. 3.2 Sign or not sign?

but difficult distinction, because all reasoning could be misinterpreted as a sign of something. Deep reflection, he says, is needed to decide what is or is not a sign. One might be dreaming (literally or perhaps day-dreaming), in what Peirce calls a "feeling" state, and a sound might evoke a reaction at a purely emotional level; this is not a sign. But, if in a state of intellectual deliberation, the same sound is encountered, its meaning is as that of a sign. In his writing Peirce refers to a steam whistle, which to the feeling state might represent some sort of existential alarm, but in a general state is literally a sign of the imminent presence of a train, a ship, or a shift change at the local factory.

Peirce is convinced that all of perception operates somewhere on this dynamical sequence, or we might call it a trajectory, of signs. A group of signs that comes to have cultural meaning, is what Peirce (1998, 10) calls a symbol. And he says, "symbols grow" and "sprea[d] among the peopl[e], in use and in experience … meaning grows." Symbols contain the prevailing character of everything; the atomic components of symbols are signs. Peirce writes compellingly of what he calls "the ten main trichotomies of signs," which are dynamical ways of interpreting the action of signs. The triangular process represented in Fig. 3.1, for instance, is the fifth of these ten trichotomies. In the end, and this is important for the domain of knowledge organization because it has direct bearing on the notion of concept-theoretic, everything may be represented with three categories: firstness (which is feeling), secondness (which is reaction), or thirdness (which is representation). The triadic sign has the quality of thirdness, as do all symbols. Pre-signs (this is my term, not Peirce's) then, are either simple emotions or emotional reactions that pre-figure signs. True categories, which would represent true concepts, must have the character of thirdness.

3.2.3   The Use of Semiotic in Knowledge Organization

It is apparent from even this brief introduction that there is potential explanatory power in semiotic theory. If knowledge organization is the science of concept-theoretic then the natural first question is how to designate the essential concepts. Semiotic theory demonstrates the difficulty in the designation of concepts, but it also demonstrates a useful approach to decision-making. From Saussure we can develop an understanding of the linguistic properties of signs, and learn to comprehend our concepts as both signified and signifier. In this way we learn that there is arbitrariness in the connection between things and their names, and there is linguistic (which is essentially social or cultural) mutability in the names of things. Smiraglia (2001a) used this distinction to explain the bibliographic entity, which has both abstract intellectual content and concrete semantic content.

More use has been made of Peirce's semiotic theory. Smiraglia (2001a, b) used Peirce's semiotic triad to demonstrate the dynamical nature of works as cultural icons. Mai and Friedman both used Peirce's semiotic theory to analyze foundational concepts in knowledge organization. Mai (2000, 2001) demonstrated the manner in which indexing as a process can be modeled using Peirce's dynamic trichotomy. The problem of interindexer inconsistency is immense in knowledge organization; Mai suggests a potential solution is to employ semiotics in the analysis of documents and the assignment of descriptors. In this manner the motion from *interpretant* to *object/interpretant* mirrors the dynamics of signs and helps indexers align their decisions with potential users of their indexing. Friedman (Friedman 2008) used both Saussure and Peirce as lenses through which to analyze the concept maps (and therefore the concepts in them) in all knowledge organization conference proceedings. In an earlier paper he had found little reference to Peircian thirdness (Friedman 2006), finding instead that concept maps tended to include first-order indications rather than complex signs, and the later, larger study confirmed this finding. Later Friedman and Smiraglia (2013) returned to the use of concept maps in knowledge organization to demonstrate the socially-negotiated identity of concepts, which are used to convey core values across time. They demonstrated a semiotic method for analysis of concept maps in which "nodes" were identified as anchors of conceptual clusters, and "arcs" between the nodes identified verbal relationship indicators. Other examples of appeals to Peirce for knowledge organization include Theleffsen (2000), Theleffsen and Theleffsen (2004), and Friedman and Theleffsen (2011). Theleffsen and Theleffsen (2004) demonstrate how semiotic theory generates a pragmatic methodology by which the essential knowledge organization schema of a domain can be shown to represent the *telos* of the domain—in essence its firstness, which includes not only its essential concepts but also its values—and thus subsequently allows the domain to move toward thirdness in the profile of the domain. Friedman and Theleffsen (2011) correlate Dahlberg's concept-theoretic for incorporating concepts in a KOS with semiotics as a philosophical basis for knowledge representation. In sum, the two are inseparable—every concept is, in a way, a sign. The role played by concepts as signs in an elementary theory of knowledge interaction is extended by Smiraglia and van den Heuvel (2013).

3.3 What Is Order? Foucault

There are two components to knowledge organization, obviously, which are "knowledge" and "organization." By "organization" we imply "order" or "sequence." Therefore while we are much concerned with questions about knowledge and how it functions, we also are much concerned with questions about order. Clearly, in the history of humankind, order usually is imposed on things and is one aspect of giving identity to phenomena. Physical sciences tend to suggest a natural order of phenomena, with sequence being one component of their syndetic nature (their connectedness). We will come back to these problems when we discuss taxonomy, because one part of the science of knowledge organization is the concordance of conceptual entities represented in different taxonomic indications, which tend to differ dramatically even among closely related domains.

But there also is thought to be a philosophical connection that exists between a language and the knowledge it represents, such that the two—knowledge and language—are interwoven. Culture, obviously, plays a large role in this interweaving because it represents common understanding that allows knowledge to be largely inferential within a cultural domain. Structure, which is closely related to sequence, and thus a constituent of order, is also part of the connection between knowledge and language. Structure designates the visible thus enabling language to facilitate communication.

Michel Foucault's *The order of things: an archaeology of the human sciences* is an existential attempt to relate the act of classification, by which order is imposed, to the cultural action of discourse, by which language mediates knowledge. Foucault thus suggests what he calls the archaeology of knowledge through discourse about the conceptions of "other" and "the same." Foucault begins by demonstrating the power of convenience, emulation, analogy, and sympathy as the typology of resemblance that constituted much of the semantic understanding before the end of the sixteenth century, when the introduction of positivist approaches began to lead away from rationalism and toward controlled empiricism. Order, he says, is about the representation of established discontinuity. Classifying in positivist natural history, he says is about the attempt to establish continuity. Thus we have set before us the components of a discourse between continuity and discontinuity, which resembles the most essential ontological question. By the end of the modern period, before deconstruction became *de rigueur*, we had returned to reliance on texts (philology he says) as a source of transcendental eschatology. In the end, human culture creates for itself a double reality in which thought and unthought must coexist—in which other and same are, essentially, coupled and therefore indistinguishable except through human discourse. Thus we find ourselves on the threshold of the post-modern era, in which thought can yield resemblance only within the visible parameters of an immediate domain.

Quinn (1994), Beghtol (1998) and Hjørland and Albrechtsen (1999) all called for essentially post-modern approaches to classification, turning away from the inflexibility of discipline-based universal schemes and toward domain-specific, multi-disciplinary, socially relevant designs. A famous paper by Mai (1999) used the term post-modern specifically to describe this movement, suggesting the abandonment of the attempt to find universal solutions. Smiraglia (2003) used this post-modern lens to deconstruct heretofore interwoven patterns of knowledge organization. All of these authors with post-modern points of view are aligned with (although none of them cite) Foucault's deconstructionist thought.

3.4   What Is a Thing: Husserl and Phenomenology

We continue with a look at Edmund Husserl's twentieth-century attempt to renew and revise Cartesian philosophy by laying out an approach to phenomenology. For Husserl, every "thing" is to be positioned over and against psychologism, which means, everything for Husserl exists only in relation to the ego. Essentially, Husserl suggests (in alignment with Peirce's semiotic theory) that each perception is subject to the interpretation of the individual. Where Husserl differs from Peirce is in the suggestion that the process of perception is viewed through the lens of personal experience.

Noesis is Husserl's perceptual component of analysis. In a series of lectures delivered at the Sorbonne in February 1929 (the content of which was later published as his *Cartesian Meditations*), Husserl (1950 [1999]) developed his notion of transcendental phenomenology. For Husserl, all perception stems from the *ego*, which is all that is. In the beginning of perception, nothing is, except that which is perceived by the *ego*. The method of perception entails a sequence of *epoche*, brackets around specific entities in the perception of the *ego*. The *epoche* is the method by which one might apprehend oneself by bracketing oneself over against the contextual world. Any spatiotemporal thing that belongs to the world exists for *ego* if it is perceived by *ego* (Husserl 1950 [1999], 21). *Eidetic description* is a process that isolates specific entities for analysis by transferring empirical descriptions into the dimension of perception (Husserl 1950 [1999], 69). Each isolate consists of its experienced form (*cogito*) and its concrete form (*cogitatum*). Perception takes place in a sequence of temporal acts (*cogitationes*). Experience, then, is a matter of the synthesis of syntheses. One sees many things at once and it is their contextual synthesis that becomes present reality. Each glimpse of the world reveals a collectivity of isolates (*cogitos*) that is perceived as the collectivity of a sequence of glimpses (*cogitationes*), each of which leads to its own *eidetic* process as well. *Noesis* comes into play at each *eidetic* moment, when we bracket an isolate to analyze it. The analysis is *noesis* and the analysand is *noema*. *Noesis* is the busying of the *ego* into whose vision the *noema* enters. Every isolate is comprehended unconsciously as an element of a larger scenario, all of which have meaning against the personal experience (*ego* acts) of the individual who is perceiving.

Husserl's traditional example is an apple tree, to which we can point or of which we can make a photograph. The particular tree is bracketed by its perception, which in turn is filtered through experience and reflected through ego. Thus the tree can be perceived as a beautiful, living, physical, pastoral entity, or it can be perceived by its noema—shade, apples, birds singing in upper branches, etc.—or, conversely, a place for daring children to swing, to climb, to scrape knees, to make a mess, and so forth. *Eideia* are like signs because they require methodological identification beyond their simple names.

3.5 And Furthermore: Wittgenstein

Which brings us to Wittgenstein and his notion of propositional signs. Wittgenstein is a philosopher who has excited scholars in knowledge organization for decades, perhaps because of the elegance of his logical process. A student of engineering and mathematics, his attempts to discover whether mathematics possessed truth, led to his work on the foundations of logic (Pears 2003, 812). One of the hallmarks of Wittgenstein's work, and especially of his second philosophical period, is his manner of deconstruction, his ability to find the particular in the general (Pears 2003, 813). For knowledge organization it is in Wittgenstein's work on the nature of language and of propositions that we find a useful approach to the meaning of a concept. For Wittgenstein it has to do with deliberation; the process is the journey. A propositional sign is a thought that has been thought out. A thought is a proposition that has sense, and the totality of all propositions is language. Propositions are themselves combinations of symbols, which are semantic "names." Names are relevant only in the context of that which they represent, or we can say, only when they are connected to that which they represent. Placing the idea of a sign in a semantic context is an approach quite similar to semiotic theory; seeing the semiotic as a semantic dichotomy is clearly parallel to Saussurian semiology.

3.6 Perception Roots the Conceptual World

Our chief concern in knowledge organization is with the identification of concepts that collectively, in some way, represent the totality of that which is known. A second critical concern is with the organization of these elements, especially with their potential conceptual orderings. These two critical questions—what is knowledge?, and what is order?—and the attendant overriding question—what is the order of knowledge?—lead naturally to the integration of philosophical approaches. As we have seen, many philosophical approaches have been borrowed by scholars in knowledge organization and these have been put to use in interesting and sometimes novel ways.

Primarily, theories about the designation of concepts and about the function of concepts as (or as parallels to) signs, leads us to semiotic theory. We look both at the dichotomous approach of Saussure's linguistic semiology and at the triadic approach of Peirce's semiosis. We find useful explanations in both approaches, and we comprehend the parallel of Wittgenstein's propositional logic very appealing because of the way it roots the concept of signs in the semantic world. We also find Husserl's approach to phenomenology appealing because of the manner in which it allows us not only to comprehend, but also potentially to quantify, perception.

In addition to our consideration of order as a phenomenon that embraces concepts, we also are released by Foucault and his contemporaries to embrace a postmodern conception of knowledge and its orders. Whereas once our domain was preoccupied with competing attempts to generate universal knowledge schemas, now we are liberated to approach individual domains. Epistemologically we have trodden a path from rationalism to empiricism and along the way we have embraced greater epistemic depth. We will discuss domain analysis specifically in Chap. 10, but it is epistemology that has led us to greater perception of multiple if diverse domains as we have turned from the fruitless search for a universal classification to the more satisfying approach to comprehension and interoperability.

References

Beghtol, Claire. 1998. Knowledge domains: multidisciplinary and bibliographical classification systems. *Knowledge organization* 25: 1–12.

Eco, Umberto. 1976. *A theory of semiotics.* Bloomington: Indiana University Press.

Eco, Umberto. 1984. *Semiotics and the philosophy of language.* London: Macmillan.

Friedman, Alon. 2006. Concept mapping as a measurable sign. In Budin, Gerhard, Christian Swertz and Konstantin Mitgutsch, eds. *Knowledge organization for a global learning society: Proceedings of the Ninth International ISKO Conference, Vienna, Austria, July 4–7, 2006.* Advances in knowledge organization 10. Würzburg: Ergon Verlag, pp. 131–9.

Friedman, Alon. 2008. Concept map as "sign;" concept mapping in knowledge organization through a semiotics lens. Ph.D. dissertation, Long Island University.

Friedman, Alon, and Martin Thellefsen. 2011. Concept theory and semiotics in knowledge organization. *Journal of documentation* 67: 644–74.

Friedman, Alon, and Richard P. Smiraglia. 2013. Nodes and arcs: concept map, semiotics, and knowledge organization. *Journal of documentation* 69: 27–48.

Goehr, Lydia. 1992. *The Imaginary museum of musical works: an essay in the philosophy of music.* Oxford: Clarendon.

Grayling, A.C. 2003. Epistemology. In Bunnin, Nicholas, and James E.P. Tsui, eds. *The Blackwell companion to philosophy*, 2d ed. Malden, Blackwell, pp. 37–60.

Hjørland, Birger. 1998. Theory and metatheory of information science: a new interpretation. *Journal of documentation* 54: 606–21.

Hjørland, Birger. 2008. What is knowledge organization (ko)? *Knowledge organization* 35: 86–101.

Hjørland, Birger, and Hanne Albrechtsen. 1999. An analysis of some trends in classification research. *Knowledge organization* 26: 131–39.

Husserl, Edmund. [1950] 1999. *Cartesian meditations: an introduction to phenomenology*, tr. By Dorion Cairns. Dordrecht: Kluwer.

Ibekwe-SanJaun, Fidelia, and Thomas M. Dousa, eds. 2013. *Theories of information, communication and knowledge: a multidisciplinary approach*. Studies in history and philosophy of science. Dordrecht: Springer.

Mai, Jens-Erik. 1999. A postmodern theory of knowledge organization. In Woods, Larry ed., *Proceedings of the Annual Meeting of the American Society for Information Science October 31, 1999, Vol. 62*. Medford: Information Today, pp. 547–56.

Mai, Jens-Erik. 2000. The subject indexing process: an investigation of problems in knowledge representation. Ph.D. dissertation, University of Texas at Austin.

Mai, Jens-Erik. 2001. Semiotics and indexing: an analysis of the subject indexing process. *Journal of documentation* 57: 591–622.

Malmkjær, Kirsten. 2004. *The linguistics encyclopedia*, 2d ed. London: Routledge.

Morrisey, Frances. 2002. Introduction to a semiotic of scientific meaning, and its implications for access to scientific works on The Web. *Cataloging & classification quarterly* 33 nos. 3/4: 67–98.

Nattiez, Jean-Jacques. 1990. *Music and discourse: toward a semiology of music*, trans. by Carolyn Abbate. Princeton, N.J.: Princeton Univ. Pr.

Olson, Hope A.. 1996. Dewey thinks therefore he is: The epistemic stance of Dewey and DDC. In Green, Rebecca, ed. *Knowledge organization and change: Proceedings of the Fourth International ISKO Conference, Washington, DC, July 15–18, 1996*. Advances in knowledge organization 5. Frankfurt/Main: Indeks Verlag, pp. 302–3.

Pears, David. 2003. Wittgenstein. In Bunnin, Nicholas and James E.P. Tsui eds. *The Blackwell companion to philosophy*, 2d ed. Malden, Blackwell, pp. 811 26.

Peirce, Charles Sanders. 1991. *Peirce on signs: writings on semiotic*, ed. by James Hoopes. Bloomington: Indiana Univ. Pr.

Peirce, Charles Sanders. 1998. *The essential Peirce: selected philosophical writings volume 2 (1893–1913)*, ed. by the Peirce Edition Project, Nathan Houser [et al.]. Bloomington: Indiana Univ. Pr.

Poli, Roberto. 1996. Ontology for knowledge organization. In Green, Rebecca, ed. *Knowledge organization and change: Proceedings of the Fourth International ISKO Conference, Washington, DC, July 15–18, 1996*. Advances in knowledge organization 5. Frankfurt/Main: Indeks Verlag, pp. 313–19.

Quinn, Brian. 1994. Recent theoretical approaches in classification and indexing. *Knowledge organization* 21: 140–7.

Saussure, Ferdinand de. 1959. *Course in general linguistics*, ed. by Charles Bally and Albert Sechehaye; in collaboration with Albert Riedlinger; trans., with an introd. and notes by Wade Baskin. New York: McGraw-Hill.

Smiraglia, Richard P. 2000. Words and works; signs, symbols and canons: the epistemology of the work. In Beghtol, Clare, Lynne C. Howarth, and Nancy J. Williamson, eds., *Dynamism and stability in knowledge organization: Proceedings of the Sixth International ISKO Conference, 10–13 July 2000, Toronto, Canada*. Advances in Knowledge Organization v. 7. Würzburg: Ergon Verlag, pp. 295–300.

Smiraglia, Richard P. 2001a. *The nature of "a work:" implications for the organization of knowledge*. Lanham, Md: Scarecrow Press.

Smiraglia, Richard P. 2001b. "Works as signs, symbols, and canons: the epistemology of the work." *Knowledge organization* 28: 192–202.

Smiraglia, Richard P. 2002a. Musical works and information retrieval. *Notes: the quarterly journal of the Music Library Assn.* 58: 747–64.

Smiraglia, Richard P. 2002b. Bridget's Revelations, William of Ockham's Tractatus, and Doctrine and Covenants: qualitative analysis and epistemological perspectives on theological works. *Cataloging & classification quarterly* 33 nos. 3/4: 225–51.

Smiraglia, Richard P. 2003. The history of 'the work' in the modern catalog. *Cataloging & classification quarterly* 35 nos. 3/4: 553–67.

Smiraglia, Richard P. 2004. Knowledge sharing and content genealogy: extending the "works" model as a metaphor for non-documentary artifacts with case studies of Etruscan artifacts. In McIlwaine, Ia C., ed. *Knowledge Organization and the global information society; Proceedings of the Eighth International ISKO Conference 13–16 July London UK.* Advances in knowledge organization 9. Würzburg: Ergon Verlag, pp. 309–14.

Smiraglia, Richard P. 2005a. Content metadata: an analysis of Etruscan artifacts in a museum of archeology. *Cataloging & classification quarterly* 40 nos. 3/4: 135–51.

Smiraglia, Richard P. 2005b. Instantiation: toward a theory. In Vaughan, Liwen, ed. *Data, information, and knowledge in a networked world: Proceedings of the Canadian Association for Information Science annual conference June 2–4 2005.* http://www.cais-acsi.ca/search. asp?year=2005 (accessed June 2014).

Smiraglia, Richard P. 2006. Empiricism as the basis for metadata categorization: expanding the case for instantiation with archival documents. In Budin, Gerhard, Christian Swertz and Konstantin Mitgutsch, eds., *Knowledge organization and the global learning society; Proceedings of the 9th ISKO International Conference, Vienna, July 4–7 2006.* Advances in knowledge organization 10. Würzburg: Ergon Verlag, pp. 383–88.

Smiraglia, Richard P. 2008. A meta-analysis of instantiation as a phenomenon of information objects. *Culture del testo e del documento* 9: 5–22.

Smiraglia, Richard P., and Hur-Li Lee, eds. 2012. *Cultural frames of knowledge.* Würzburg: Ergon Verlag.

Smiraglia, Richard P., and Charles van den Heuvel. 2013. Classifications and concepts: towards an elementary theory of knowledge interaction. *Journal of documentation* 69: 360–83.

Theleffsen, Thorkild. 2000. Firstness and thirdness displacement: epistemology of Peirce's sign trichotomies. *AS/SA* 10: 537–52.

Thellefsen, Martin and Thellefsen, Torkild. 2004. Pragmatic semiotics and knowledge organization. *Knowledge organization* 31: 177–87.

Thomas, David H., and Richard P. Smiraglia. 1998. Beyond the score. *Notes: The quarterly journal of the Music Library Association* 54: 649–66.

Chapter 4
History: From Bibliographic Control to Knowledge Organization

4.1 A Social Confluence at the Center

Throughout the history of what we now call knowledge organization there is woven a single, central thread, which will be the focus of this brief chapter. This thread is the confluence of art, commerce, and technology. As you will see, most of the history of the organization of recorded knowledge is rather colorless, which is a nice way of saying there is nothing much to report. A small number of documents existed, everybody who needed to know them did so, if anybody wanted to know about them one of the experts could be questioned, and that was it. Substantial change takes place historically when these three social threads—art, commerce, and technology—come together at important moments to act as a collective catalyst to move the domain forward. This thesis was first put forward in music bibliography (Young 1982) in a remarkable piece about the growth of music printing. But as we shall see, it is true not just of significant moments in printing or in bibliography but in the control of knowledge as well.

A second major thesis is that the evolution of knowledge organization roughly parallels the development of democratic societies. As the growth of societies dependent on a voting public demanded greater public education, the need to know was accompanied by the need for ever better tools for knowing what it is that any given society knew. Thus, the growth of an educated populace led not only to more education, to more educational institutions, and to greater literacy, but also to more sophisticated tools for the ordering of knowledge.

A third thesis, which like the first will permeate this narrative, will be that there have been more, and more rapid, developments in the past century than in the entirety of time before. The distance between developments of importance in the organization of knowledge was millennia followed by centuries, until (arguably) the mid-twentieth century witnessed the ramping up of trends that find their early threads in the mid-eighteenth and mid-nineteenth centuries. But it was

© Springer International Publishing Switzerland 2014   33
R.P. Smiraglia, *The Elements of Knowledge Organization*,
DOI 10.1007/978-3-319-09357-4_4

developments from the period of the First World War up to the present that are most closely parallel to the greatest innovations in the control of recorded knowledge.

In this short chapter we will look separately at each of these theses, and then we will look in greater detail at the development of the domain now known as "knowledge organization." Despite the danger inherent in attempting to chronicle the history of the present, it is important to grasp both the obvious chronological imperative of the development of the science of concept-theoretic and this crossroads in the history of information, as well as the important distinction between that science and all that preceded it.

4.2 The Chronology of Bibliographic Control

Readers are referred to the grand and very well-known narrative of Ruth French Strout (1956) for the entertaining details of the history of catalogs and cataloging. For a narrative history of the development of schools of thought readers are referred to Collins (1998). There you will discover the social realities of the growth of knowledge over the whole course of human history. The myth that knowledge is subject to magical discovery on special occasions dictated by chance is put to rest in Collins' narrative of the social forces that impel the growth and suppression of ideas. Rather than revisit either of these narratives, we will look at a few important historical developments that demonstrate our thesis about the social confluences that push the organization of recorded knowledge forward.

4.2.1 Antiquity—Lists

An overview of the chronology of the control of recorded knowledge also is revealing concerning how little we know. We know that lists of books survive from antiquity and we know that there were libraries in antiquity, but we do not know whether any of the lists were actually catalogs. Callimachus, the librarian at Alexandria compiled a list, referred to as "Pinakes" (it is Greek for "board" or "tablet") (Strout 1956, 256 ff.). Not only was this a list of works in his collection, but also it was an ordered list. Callimichus' list was organized in broad subject categories, with various sub-arrangements in each, and with individual entries identified by their physical location and title or opening words. What we see here, as in other ancient sources, is some sort of intellectual compilation that gathers material by subject, and with some sort of functional prerogative. We cannot know how these lists might have been used or why they might have been made. So we can refer to antiquity with a shrug of our intellectual shoulders, wishing we knew more about their bibliographical practices. But we cannot really point to that period for guidance about our own practices. What we see is bibliographical scholarship that is in alignment with other scholarship of the time.

4.2.2   Middle Ages—Inventories

Our next stop will be the middle ages in Europe and even then there is not much new to report. By this time monasteries had become principle repositories of books, probably because the Christian clergy were literate and what scholarship existed at the time took place mostly among them. A simple fact for the twenty-first century mind to grasp is that to make a book in this time (indeed, as at all times prior to this) meant to write it out by hand. To acquire a copy meant similarly to cause a copy to be written by hand. Books were precious and rare by their very nature, and therefore they also were likely to be known well by those who possessed them. The very modern idea of a large building filled with mysterious tomes in need of mechanized indexes is in no way applicable to this period. Instead, we continue to see lists as before and these continue to be in accord with the scholarship of the day, which as Strout points, out was primarily involved with compilation (Strout 1956, 258). By the ninth century it was clear that monastery library inventories were being compiled and kept, but for what use we do not know. Collections contained a couple of hundred volumes of works bound together. The inventories were for the purpose of accounting to various authorities for the contents of the institution.

4.2.3   Seventeenth Century—Finding Aids

Our real journey toward the modern catalog begins in the seventeenth century with the introduction of functional finding lists. By the sixteenth century bibliographers had begun to compile more sophisticated lists of books, and it was the Zurich bibliographer Konrad Gesner whose 1548 *Pandectarum* (Strout 1956, 263) introduced the "see" reference, and standardized forms of classification. By 1595 Andrew Maunsell had produced his famous *Catalogue of English Printed Books*, which featured entry of authors under surnames, anonymous works under title, uniform locations for editions of the *Bible*, and added entries for subjects and translators. This remarkable advance begins to look a lot like the modern catalog, although we must remember that it still was in book form and still was a catalog of books for sale rather than an index of the contents of a library. Let us reflect now that it was around 1439 that Gutenberg created the mechanisms for printing from movable type that were to revolutionize the printing of books. We are looking, then, at the flowering of the marketplace for books only a bit more than a century after this remarkable invention. It was the need of the marketplace that drove the development of more sophisticated forms of bibliography. It is from this era that the notion of the encyclopedia as the unified record of knowledge emerges from scholars. The Nuremberg Chronicle—officially the *Liber Chronicarum*, written by Hartmann Schedel and printed in 1493—was an early example of such a volume. A century later, according to Strout (1956, 264) the notable scholars of the age—Descartes, Bacon, Galileo—were the generators of ideas concerning the order of knowledge as discerned by scholarship.

4.2.4   Nineteenth Century—Collocating Devices

Our story really picks up with Sir Thomas Bodley, who together with librarian Thomas James, built a new library for Oxford University in the seventeenth century. Because the acquisition of books was a critical functional imperative of their bibliographical tool, we now see some of the practices originated by Gesner and Maunsell appearing for the first time in the catalog of a library. The arrangement was classified but with an alphabetical author index by surname, and it included analytical entries (for the individual contents of volumes). We had entered the era of the finding list, at last an improvement from the inventories and lists of the past.

The nineteenth century was the beginning of the period of modernity for the library catalog, and we see parallel developments in continental Europe, Great Britain, and the (now) new American scholarly communities. In relatively rapid succession we encounter the cataloging rules of the revolutionary government in France (1791), with its instructions for brevity and for the use of cards, Antonio Panizzi's catalog for the British Museum Library (1839), in which the imperative of the collocation of works is clearly a high priority, and the evolution in the United States of cataloging codes for the Smithsonian (1850) by Charles Coffin Jewett, and by 1876 Charles Ammi Cutter's famous *Rules For a Printed Dictionary Catalog*. Two developments deserve notice here. First, that we now have moved into the territory of libraries as repositories of the record of knowledge and, therefore, as its curators. And second, that the catalogs now being developed are sophisticated collocating devices, stipulating entry for each work (not book, but work) under author, title, and subject, and in some cases under classification as well. With these tools we see a shift from the scholar to the library in responsibility for the universal index of recorded knowledge.

Cutter's introduction sets out these "objects" (we would say "objectives" today) for the catalog (Cutter 1876, 12):

1. To enable a person to find a book of which either

   (A) the author ⎫

   (B) the title ⎬ is known.

   (C) the subject ⎭

2. To show what the library has

   (D) by a given author
   (E) on a given subject
   (F) in a given kind of literature.

3. To assist in the choice of a book

   (G) as to its edition (bibliographical).
   (H) as to its character (literary or topical).

These objectives (and their recitation) are ubiquitous in cataloging scholarship for they have shaped the structure of catalogs around the world from Cutter's time to the present. Although mechanization of catalogs led rapidly from printed book catalogs of the late nineteenth century to the card catalogs of the first three quarters of the twentieth century to the online catalogs that emerged in the last quarter of the twentieth century, the structure of the catalog has always paid homage to this list of objectives. It means that for every work, the catalog must facilitate the location (or finding, or identification) of an author, a title, or a subject, when any one of the three is known. It means that the catalog is structured not only to provide recall for known-item searches, but also to allow browsing in collocated lists (collocation means literally to place entries side-by-side; it is the gathering function of the catalog) of author's names, subject terms, and literary genres. And finally, the catalog is intended to contain enough information about each book to allow the user to make an informed choice from among the many that might be collocated. Whether these objectives function well or not will be our subject in subsequent chapters. But the impact of Cutter's objects on the history of the organization of knowledge cannot be emphasized enough. It is the primary benchmark by which catalogs were developed and maintained for a century and a half. It is only in the present time that the sequence of objectives, or the addition of new objectives, has even arisen as a subject of controversy in the professional community.

4.2.5 Twentieth Century—Codification and Mechanization

The twentieth century was a time of mechanization and codification. The growth of professional associations of librarians, especially the partnership that developed between The Library Association (of Great Britain) and The American Library Association, led to increased international standardization of cataloging practices, and in particular, of rules for the construction of catalogs. All of these rules continued to be based on Cutter's objects to varying degrees, but with rapid development of detailed instructions, particularly for the representation of specific sorts of bibliographic metadata. This was just one factor that brought about the famous mid-twentieth century "crisis in cataloging" (Osborn 1941; Yee 1987), which threatened to overwhelm the American bibliographical apparatus. Osborn's famous appeal to pragmatism was really a call for greater bibliographical professionalism, and is presented over and against three less appealing professional metaphors. These were the perfectionist, striving for every jot and tittle to the point that nothing ever got finished, the legalist who could make no decision without appeal to experts and the codified rules, and the bibliographer who could not ignore any change in font or paper quality in the description of a text. The pragmatist was an iconic professional who could move a pile of books swiftly through the cataloging process making them available for use by scholars.

The crisis noted by Osborn was an immense backlog of uncataloged books in the Library of Congress, but it was part of an even larger problem at mid-century. That was the inability of the entire bibliographical apparatus to keep up with the production of new knowledge. This became a crucial strategic problem during the period of World War II, when the western Allies sought to win not just an armed conflict but also a scientific contest to realize nuclear ambitions before the Axis powers could. It was this period that began the merger of mechanical bibliography that was the product of the documentation movement with the highly codified bibliographical practices of librarianship, which merger would lead in the 1960s to the development of the American Society for Information Science. Nevertheless, the experimentation with new mechanical devices for helping scholars rapidly to index and retrieve scientific data can be seen alongside the streamlining of procedures for bibliographical resource description.

By 1950, the war was won and ended, the world was pondering reconstruction and a new world order, and the famous conference on bibliographical organization held at the University of Chicago's Graduate Library School (Shera and Egan 1950) became a hallmark of visionary thinking that would lead directly to the development of the field of knowledge organization as we know it today. Shera and Egan opened the conference (and the volume) with their "Prologomena" that equates bibliographic control with documentation, and calls for the development of devices that can direct intellectual energy in extracting information relevant to specific tasks (Shera and Egan 1950, 17). They pose the notion of a contrast between "internal" bibliographic control, which exists in libraries, and "external" control, which is used to fuel intellectual activity in general, thus foreshadowing Wilson's notion of descriptive and exploitative domains. They describe the publishing growth crisis of the first half of the twentieth century and the attendant loss by libraries of the control of indexing of scientific journals, which became a highly profitable private sector enterprise. Perhaps the most influential papers in the volume are those by Clapp and Shera. Clapp (1950, 4) describes the social role of bibliographic organization as one of the arts of communication, and lays out quantitatively the dimensions of the challenge facing the bibliographical, library, and documentation communities at mid-century. Classification was clearly thought to be a critical requirement for resolving the crisis and both Shera and Ranganathan were present to suggest approaches. Shera called for research that could reveal the essential conceptual structure of human knowledge (Shera 1950, 72).

In 1961 the historic post-war International Conference on Cataloging Principles held in Paris produced the first re-working of Cutter's objects in a century, and led to international agreements on bibliographic principles and practices that would fuel the increasing conformity of cataloging rules for the remainder of the twentieth century. The last quarter of the twentieth century saw the introduction of computerized online public access catalogs, or OPACs, that mechanized the card catalogs produced in the first part of the century. Nicholson Baker famously (and arguably correctly) decried the discarding of these catalogs by librarians as shortsighted (Baker 1994). While librarians could hardly be criticized for discarding the dinosaurs in favor of the space-age online catalogs, it is true that the old card catalogs represented immense feats of engineering. Like the Golden Gate Bridge or the Eifel Tower, the card catalogs of the world's major libraries were a testimony to human effort and to the desire to organize recorded knowledge, to harness knowledge for human purposes. One hopes some catalogs have been kept as artifacts for scholars of later centuries to study, to provide a glimpse of our own time (rather than leaving no trace of centuries of bibliographical engineering, and rendering history uninformed, rather as we have no such picture of antiquity's bibliographical practices).

4.3 The Rise of Public Education

As we have seen, the evolution of knowledge organization has several parallels, including the confluence of technology that was the beginning of printing from movable type, the information revolution of the twentieth century made possible with the introduction of even faster printing technologies, and the digitization of knowledge at the end of the twentieth century. There also are parallels, less marked, with the development of scholarship itself, and notably, some of the most important innovations in the design of knowledge come from scholars themselves rather than from information professionals. But as we suggested earlier the rapid advance of the collocating device also, arguably, parallels the development of democratic societies. As the growth of societies dependent on a voting public demanded greater public education, the need to know was accompanied by the need for ever better tools for knowing what it is that a society knew. Thus, the growth of an educated populace led not only to more education, to more educational institutions, and to greater literacy, but also to more sophisticated tools for the ordering of knowledge.

We have already seen the influence of the French Revolution on worldwide practices of cataloging. According to Smalley (1991), the introduction of these rules not only represented the first national approach to organizing recorded knowledge, it also represented the first set of carefully constructed pragmatic instructions for the organization of knowledge. Never mind that the purpose was to facilitate sales of the books confiscated by the provisional government, or that their scheme never paid off. The rules themselves are a hallmark of pragmatic thought.

We can also observe the parallel between the rise of democratic society in both the United States and in Great Britain. No less than the 1876 report containing Cutter's rules is our starting point. That report was just an appendix to a major report commissioned by the United States Bureau of Education on the role and status of the public library. Issued at a time when sixth grade education, if any, was a norm and literacy was not by any means a national priority, the public library was being held up as a way for local governments to maintain programs of lifelong learning for their citizens. It was this notion that led to the standardization of library practices. This not only fueled the growth of librarianship as a profession but it led to major economies of scale in the bibliographic industry, such as the introduction of printed catalog cards, sold for a pittance, by the Library of Congress. By the mid-twentieth century the majority of cataloging in U.S. libraries was coming from the Library of Congress and its participating partners. This freed librarians from the chores of bibliography and left them time for tending to the social needs of their communities. Library service was both a public and cultural role, and the collocation of books to facilitate browsing was an important aspect of it. No longer warehouses for unused books, libraries became the working person's working collection— cookbooks, mechanical guides, planting and harvesting pamphlets, medical advice—all of this became centralized and available at the local public library. And all of it had to be controlled intellectually.

By the middle of the twentieth century Clapp (1950) was writing that bibliography was one of the arts of communication found at a second level of utterance, treating prior records of communication, and in need of patterns of effective arrangement. Such arrangements were no less than a matter of national security intended to lead the inquirer to critical new communications. He wrote (1950, 4 ff.):

> The problems of bibliography may seem at any one moment to predominate over its achievements; but is accomplishments are majestic. Without bibliography the records of civilization would be an uncharted chaos of miscellaneous contributions to knowledge, unorganized and inapplicable to human needs.

In the same volume, Jesse Shera and Margaret Egan referred to social role of bibliography as part of the problem of inter and between group communication (1950, 17). Collins has written that (1998, 429) "the social process of importing ideas constructs the meaning of what is being conveyed." The means by which ideas are moved from community to community, from scholar to scholar, from individual to individual and across all of these as well, has the effect of stimulating new knowledge itself.

4.4   The Discipline: Knowledge Organization

Finally, we will look briefly at the outline of the recent evolution of knowledge organization as a domain. As you have seen, the domain has roots in bibliography, in publishing, in librarianship, in scholarship, in documentation, an in several other fields as well. As we now understand it, knowledge organization is found at the intersection of information retrieval and social dynamics. It is seen as the process of structuring of knowledge that borrows from logic, psychology, linguistics, semiotics, epistemology, and informatics. We can date the intension of the domain from the first publication of its journal, then titled *International Classification,* which first appeared in 1974. Founded by Ingetraut Dahlberg, a noted German indexer, documentalist and publisher (Dahlberg 2008, 82), *International Classification* (and its successor from 1993 *Knowledge Organization*) became the primary venue for the interdisciplinary treatment of what Dahlberg called "concept-theoretic." Intended to bring knowledge representation under one roof, Dahlberg later founded the International Society for Knowledge Organization in 1989. In 1993 she created a classification system for knowledge organization literature (Dahlberg 2006, 15), which was intended to reveal both the intension and extension of the domain.

According to Dahlberg (2008, 83) the need for a concerted effort at thesaurus construction, in particular multi-lingual machine-assisted thesaurus construction, was a primary impetus for the founding of the society. Given her understanding of the role of the sciences in the pursuit of new knowledge, Dahlberg set out to give her domain all that it would need to thrive—a journal, an international society, and a theoretical base. Thus at the close of the twentieth century the history of attempts to organize recorded knowledge brought together the competing forces from scholarship, culture and society, and information processing to yield a new, multi-disciplinary science. From this point onward, our narrative will turn to attempts to define the new domain's theoretical base, which Dahlberg called "concept theoretic." What is a concept, and what is a theory, will become primary questions for knowledge organization.

References

Baker, Nicholson. 1994. Discards. *The New Yorker* 70: 64–70.
Clapp, Verner W. 1950. The role of bibliographic organization in contemporary civilization. In Shera, Jesse H. & Egan, Margaret E. eds. *Bibliographic organization: papers presented before the fifteenth annual conference of the Graduate Library School July 24–29, 1950*. Chicago: Univ. of Chicago Press, pp. 3–23.
Collins, Randall. 1998. *The sociology of philosophies: a global theory of intellectual change*. Cambridge: Belknap.
Cutter, Charles Ammi. 1876. *Rules for a printed dictionary catalog*. Special report. Part II. In United States Bureau of Education, *Public libraries in the United States*. Washington: Govt. Printing Office.
Dahlberg, Ingetraut. 2006. Knowledge organization: a new science? *Knowledge organization* 33: 11–19.
Dahlberg, Ingetraut. 2008. An interview with Ingetraut Dahlberg. *Knowledge organization* 35: 82–85.
Osborn, Andrew D. 1941. The crisis in cataloging. *Library quarterly* 11: 393–411.
Shera, Jesse H. 1950. Classification as the basis of bibliographic organization. In Shera, Jesse H., and Margaret E. Egan, eds. *Bibliographic organization: papers presented before the fifteenth annual conference of the Graduate Library School July 24–29, 1950*. Chicago: Univ. of Chicago Press, pp. 72–93.
Shera, Jesse H., and Margaret E. Egan, eds. 1950. Bibliographic organization: papers presented before the Fifteenth Annual Conference of the Graduate Library School, July 24–29, 1950. The University of Chicago studies in library science. Chicago: University of Chicago Press.
Smalley, Joseph E. 1991. The French cataloging code of 1791: a translation [issued by the revolutionary French government for cataloging libraries seized from religious houses]. *The Library Quarterly* 61: 1–14
Strout, Ruth French. 1956. The development of the catalog and of cataloging codes. *Library quarterly* 26: 254–75.
Yee, Martha. 1987. Attempts to deal with the crisis in cataloging at the Library of Congress in the 1940s. *Library quarterly* 57: 1–31.
Young, James Bradford. 1982. An account of printed music ca. 1724. *Fontes artis musicae* 29: 129–36.

Chapter 5
Ontology

5.1 Ontology Is About "Being"

Like many aspects of information, ontology is a difficult term with several meanings used in different ways in different parts of the domain. The point of departure, however, for all of these connotations is the use of ontology as a domain of thought in philosophy. In philosophy, ontology is the study of being—of what is. Point of departure is a good metaphor here, because of course many roads lead from this one point all going off in different directions. Whether they meet in Rome, as it were, or not, is another matter. For instance, the question "what is?" naturally implies the question "what is not?" The question "what is" implies the subquestion, "then, what is one?" Of course, the question of "how do I know?" also arises—that is the province of epistemology—a superhighway of its own as we saw in Chap. 2.

In philosophy, ontology allows us to isolate certain principles of physical vs. metaphysical, of categories and the entities that are their contents, of the relationships among all of the above, of attributes of phenomena such as facts, properties, energy, space, time, etc. We have already encountered some of these aspects in our discussions of semiotics and phenomenology. In particular, with Husserl and his philosophy that evolved from Descartes's famous maxim, we saw a particular metaphysical point of view based on the ego at the center of being—*Cogito ergo sum*, (*I think, therefore I am*)—meaning that for me all that exists does so only as I perceive myself standing alongside it.

It is easy to extrapolate from the preceding paragraph the manner in which ontology informs knowledge organization. If, for example, our job is to consider all of reality as though it were a bowl full of marbles, and somehow to divide the marbles into convenient and sensical bags according to some conceptual scheme, then we start with the marbles (Fig. 5.1).

Now we must ask ourselves over and over "what is?," therefore "what is not?," and then "what is one?," and therefore "what is not one?" By these questions we divide our marbles into red and blue, large and small, and then we have to decide how

© Springer International Publishing Switzerland 2014
R.P. Smiraglia, *The Elements of Knowledge Organization*,
DOI 10.1007/978-3-319-09357-4_5

Fig. 5.1 Marbles in a bowl
(Dusty 2008; http://www.
flickr.com/photos/9243200@
N04/1010233398/. Accessed
11 October 2008; image not
edited)

further to divide them as well. We want to consider the heuristics of the situation—
that is, the rules by which we make these divisions. Are we dividing our marbles
according to the inherent properties of marbles? Or do we divide them according to
rules for retrieving them according to some pragmatic user-based heuristic? Here we
have the question of epistemology—is our ontology empirical or pragmatic?

Other nuances abound, of course. Do we have two classes—red and blue? Or do
we have two classes—large and small? If we have only two classes, which is more
important, size or color? It should be clear that we do not have a proper hierarchy
here, because neither size nor color is determinative of the other. In this case we
have identified not essential ontological qualities, but rather typological qualities.
A typology simply describes its entities according to their characteristics—this is a
big red one, this is a small blue one—and so forth. Typology lends itself to the style
of organization called "faceting," from facet as a single face of a gem such as a
diamond. It takes all of the facets to describe the diamond, and all of them intersect
in the diamond, and all of them represent intersecting geometric planes metaphysi-
cally. For instance, as you can see in Fig. 5.2, a diamond has many facets (faces).

Imagine that each of those facets represents a different characteristic of a phenomenon. It is easy to see that we then would require all of the facets and their
points of connection to describe the gem. What if we break each facet out for separate analysis. We would have something like what you see in Fig. 5.3.

Now we can see how facets are generated by separating the faces—the presentation characteristics, as it were—of a phenomenon. We can further identify each
facet by supplying all of the terms and relationships among them that are appropriate. Each facet, of course, describes its own ontology.

In information the term ontology is used to describe a knowledge base. In both
computer science and information we see the construction of sets of terms
(concepts) that are used in a specific community—or domain—mapped together
with the relationships among them. The resulting map is called the ontology of
a domain. Figure 5.4 is a simple map (simplistic might be a better term) of the
banking domain.

You will notice several things about this ontology. First, because it is domain-
specific, it represents only a single component of reality. It is not everything that is

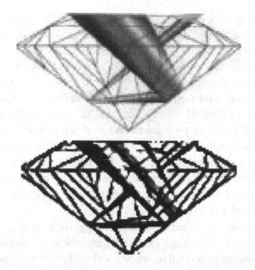

Fig. 5.2 The facets of a diamond (Newcentury 2008; Newcentury. 2008. 100facet.jpg. http://www.newcenturydiamond.com. Accessed 11 October 2008; images cropped and edited)

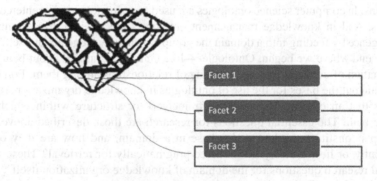

Fig. 5.3 Facets collectively describe the phenomenon

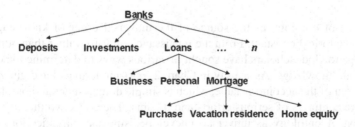

Fig. 5.4 A domain-specific ontology

known, or everything that could be known about banks; rather, it is just a set of terms that are needed by the domain with which we are working at the moment. A second concept clearly evident here is that there is only one facet and its contents are arranged hierarchically in what is called a tree-structure (a tree-from-a-point, although, admittedly, the image is inverted). Sometimes a simple hierarchical arrangement is enough. And sometimes facets are defined by different epistemologies. We have here a standard set of terms that describe kinds of bank functions, kinds of loans, and kinds of mortgages. What if we needed to represent human interaction with banking institutions? That would generate terms from a different point of view—a potentially different facet. The final observation about this simple illustration is that the ontology is incomplete (see the ellipsis and the *n*), which means new terms can be added as understanding of the domain evolves. Thus ontologies are dynamic and not fixed.

In all disciplines ontologies exist—in archeology, for instance, typologies are created to describe artifacts. These are essentially empirical ontologies. In the physical sciences, taxonomies are generated to define (or classify) entities and the relationships among them as they are discovered by empirical research. In knowledge organization ontologies are used to generate classifications and thesauri—standard knowledge organization systems (KOS). A KOS is a system that uses the inherent structure of an ontology to order the concepts for retrieval. In information science ontologies are studied for clues to the evolution of knowledge in, between, among, and across domains. In computer science ontologies are used to generate domain-centered applications. And in knowledge management ontologies are used to drive competitive intelligence by feeding into a domain the group's own knowledge about itself.

We end where we began. Ontologies—like Ontology—are all about being—the designation of concepts and the mapping of relationships among them. Poli (1996, 313) laid out the bases for the use of ontologies in knowledge organization, reminding us that an ontology is a general framework or structure within which KOS can be built. The essential questions for research are those described above. What concepts constitute the knowledge base in a domain, and how are they ordered inherently, or how are they best ordered pragmatically for retrieval? These are the critical research questions for the domain of knowledge organization itself.

5.2 Encyclopedism and Classification as Ontological Enterprise

The notion of ontology as the source of the universal order of knowledge is, of course, a historical constant. From the renaissance onward, rather than attribute all knowledge to God, scholars have sought in various ways to determine the one, true order of all knowledge. As we say in Chap. 2, post-modernists have given up that effort in lieu of the accomplishable, which is simple domain-centered ontology. But not before centuries of serious efforts and dreams. There are two threads really in the notion of ontology as universal: encyclopedism, and classification. We will describe here, briefly, some major efforts in each stream.

5.2.1 Encyclopedism

Encyclopedism is the notion that all knowledge can be brought together in one place. At one time, it was considered feasible to create a compendium of all knowledge, which would become the universal source. One of the most famous (and most frequently cited as foundational) was the effort of French philosopher Denis Diderot (1713–1784), known as *Encyclopédie, ou dictionnaire raisonné des sciences, des arts et des métiers*. Diderot spent 20 years compiling and writing his encyclopedia, and perhaps the most lasting influence, aside from his devotion, is his method. A complete philosophy underlay the entire undertaking, thus influencing every concept represented. He sought to allow lay people, for the first time, access to the most recent and useful knowledge. He brought together knowledge compiled by scholars and artists as well as trades people, the better to provide social integration of useful knowledge. Diderot's compendium was highly influential, and arguably contributed to the liberalization of knowledge that led to the French revolution. It is clearly the successor of today's tools such as *Encyclopedia Brittanica*, and in the *Brittanica* one can see clearly the outline of the classification of all knowledge that informs the compendium—tribute to Diderot's determination.

Perhaps most famous in bibliographical circles, and seen clearly as a forerunner of the Internet, are the famous Belgian team of Otlet- and LaFontaine. Paul Marie Ghislain Otlet (1868–1944) who together with Henri LaFontaine (1854–1923) envisioned a central repository of all knowledge of every kind. In 1895 they called an international conference to discuss this "repertory," which would make use of newly perceived global means of bibliography. Their efforts led to the development of a process known widely as "Universal Bibliographic Control," which was technologically much before its time. Otlet and LaFontaine called their vision the "Mundaneum" or City of Knowledge, and they were given a wing of a palace in 1924—hardly the "city" they envisioned. Eventually they did manage to compile an immense card catalog with 16 million entries made by 1930 (Rayward 2008a, 13). The Mundaneum was to have been the home for a new kind of documentation that made use of artifacts of all sorts, which would, in turn, lead to a new global information society. Otlet intended to "transform knowledge" by disassembling and reassembling the facts available in individual sources and make his newly transformed knowledge available via electronic broadcast media such as radio and television (Van den Heuvel 2008, 132).

In 1938 science fiction author H.G. Wells (1866–1946) wrote several essays about what he called the "World Brain." It was to be a digest of all human knowledge, and in particular it was intended to help scholars to synthesize disparate knowledge. He called it a "mental clearing house" for the mind. Like Diderot and Otlet before him, Wells intended to re-create society into a global "world-state." Wells' notion was to make use of social engineering, which only would work if all of the best scientific knowledge were available. Thus Wells' World Brain was to be a utopian encyclopedia-library-museum-archives-gallery-atlas etc., which would represent the "memory of Mankind" whose tentacles would spread everywhere so as to constantly incorporate new discoveries (Rayward 2008b, 233). Indexing, of course, was to be the essential linchpin, which prove to be as technologically elusive as it had been for Otlet. In the immediate post-World War II era, Viennese social scientist Otto Neurath envisioned a universal scientific community, which he referred to as "Unity of Science." Neurath's unity would bring together the physical sciences and the social sciences under the shared epistemology of logical positivism, all of which would be united with a projected encyclopedia of unified science. Neurath was particularly intrigued by the power of visual images, and his efforts to escape the limits of alphabetic speech led to inclusive forms of visual communication (Hartmann 2008, 283).

Within each of these encyclopedic enterprises lay the utopian goal of making all of the world's knowledge available not only to scholars but to all citizens of the global society as well. It is an expression of confidence in what Patrick Wilson (1968) later would refer to as exploitative bibliographic power. It is human dedication to the goal of illuminating the pathways that connect diverse bits of knowledge in order to make life better in every way imaginable. These early efforts floundered for a variety of reasons, many political (especially in the cases of Diderot and Neurath). Even more crucial, however, was the absence of sufficiently sophisticated automatic bibliographical apparatus. By the 1990s the apparatus had begun to appear, with the integration of computerized bibliographical techniques for cataloging of library collections and archival documents (Smiraglia 1990).

5.2.2 Universal Classification

Of course, the other great ontological tool is classification, or the systematic ordering of knowledge. In this case, the word "systematic" implies an underlying structure, which is imposed on the ontological conceptual content, as well as a form of symbolic representation that frees the structure from the hindrances of language. Classification was to find immense use as a bibliographical tool for the organization of libraries. Classification also is used throughout society for the ordering of all kinds of knowledge that passes through professional hands. Medicine, taxes, banking, insurance, and many, many more of the enterprises that govern our lives are themselves governed by classifications (see for example Bowker and Star 1999).

One of the most important advances in the theory of classification was introduced by Otlet as the organizing device for his Mundaneum. This is the *Universal Decimal Classification*, which began with Melvil Dewey's ten basic classes but added flexibility by allowing faceting to bring together (or synthesize) otherwise disparate concepts (this is called analytic-synthetic classification). We will revisit the UDC in our chapter on bibliographic classifications, but the reason for mentioning it here, is that it was Otlet's intention that the UDC should be the structural language of the ontology of everything. By borrowing the DDC's conceptual content, one guarantees a hierarchical arrangement of concepts that is derived from literary warrant, meaning it is derived that the literature collected by libraries. And by adding the concept of faceting to create an analytico-synthetic structure, UDC achieves (to a degree) its goal of having the capability of describing anything or everything. Today the UDC is maintained by a consortium that is a part of the legacy of Otlet; one can find them here: http://www.udcc.org/. Of course, like most classifications, the epistemology of its designers is limiting. And this always will be the fallibility of universal KOS, that every choice of "what is?" leaves out too many relevant phenomena. Ontology, ultimately, is about what *is*.

5.3 Toward Domain Analysis

This brings us naturally to the question of how ontology is to be constructed. We have already visited, in colloquial form, the key research questions. Now we turn to the tools of ontology generation. Most of these tools are empirical and document-based. Empirical approaches use the tools of both major research paradigms—logical positivism, and symbolic interaction/naturalistic inquiry—to look for the set of concepts that govern practice in a domain. There are several key problems, one of which is how to define a domain, that have no good answers. We will revisit those in a later chapter on domain analysis. For now we want simply to consider the few empirical approaches available to us for generating the concepts that constitute a domain.

The best methodologies are those that are often referred to as bibliometric (or more recently informetric). But where bibliometric techniques usually focus on identifying networks of scholarly communication, in this case we are concerned with terminology. Various software packages are employed, but the basic techniques are to sort words in titles of papers or in abstracts. The purpose of the sorting is to identify (using frequency distributions in descending order) distributions of most-used terminology. Smiraglia (2007, 2013) are examples of this technique.

Another common approach is to select representative documents, and to use these to generate conceptual terminology. An example of this approach is found in O'Keefe (2004). In this case, the index and table of contents are scanned for keywords, and then the full text is searched for the quantitative incidence of those terms.

The other major approach to ontology that is gaining currency in knowledge organization is called Cognitive Work Analysis (CWA). CWA relies on the assumption that there is shared implicit knowledge among the people who work closely together (in this case, *work* means job duty). CWA uses a schema promulgated first by Rasmussen and others (1994) to identify actors and constraints within a work domain (Mai 2008). The layers of knowledge are unpeeled through analysis—a visual of an onion frequently is used to explain this part of the methodology. Like most bibliometric methods, CWA is primarily a qualitative method, using ethnographic techniques to get inside the domain to study the ontology of the actors, but qualitative analyses also are employed. Pejtersen and Albrechtsen (2000) produced some of the first results for KO using this method to identify the ecologies of work spaces. More recently Marchese (2012) used CWA to analyze emergent knowledge in a human resources firm; Marchese and Smiraglia (2013) extended the study to demonstrate the roles of emergent vocabulary in a work environment.

References

Bowker, Geoffrey C., and Susan Leigh Star. 1999. *Sorting things out: classification and its consequences*. Cambridge: MIT Press.

Hartmann, Frank. 2008. Visualizing social facts: Otto Neurath's ISOTYPE project. In Rayward, W. Boyd ed. *European modernism and the information society: informing the present, understanding the past*. Aldershot: Ashgate, pp. 279–94.

Mai, Jens-Erik. 2008. Design and construction of controlled vocabularies: analysis of actors, domain, and constraints. *Knowledge organization* 35: 16–29.

Marchese, Christine. 2012. Impact of organizational environment on knowledge representation and use: cognitive work analysis of a management consulting firm. Ph.D. dissertation. Long Island University.

Marchese, Christine, and Richard P. Smiraglia. 2013. Boundary objects: CWA, an HR firm, and emergent vocabulary. *Knowledge organization* 40: 254–59.

Pejtersen, Annelise Mark, and Hanne Albrechtsen. 2000. Ecological work based classification schemes. In Beghtol, Clare, Lynne C. Howarth, and Nancy J. Williamson, eds. *Dynamism and stability in knowledge organization. Proceedings of the Sixth International ISKO Conference*. Advances in knowledge organization 7. Würzburg: Ergon Verlag, pp. 97–110.

Rasmussen, Jens, Pejtersen, Annelise Mark, and Goodstein, L.P. 1994. *Cognitive systems engineering*. New York: Wiley.

O'Keefe, Daniel J. 2004. Cultural literacy in a global information society-specific language: an exploratory ontological analysis utilizing comparative taxonomy. In McIlwaine, Ia, ed. *Knowledge organization and the global information society: Proceedings of the 8th International ISKO Conference, London, July 13–16, 2004*. Advances in knowledge organization, v. 9. Würzburg: Ergon Verlag, pp. 55–59.

Poli, Roberto. 1996. Ontology for knowledge organization. In Green, Rebecca, ed. *Knowledge organization and change: Proceedings of the Fourth International ISKO Conference, Washington, DC, July 15–18, 1996*. Advances in knowledge organization 5. Frankfurt/Main: Indeks Verlag, pp. 313–19.

Rayward, W. Boyd. 2008a. European modernism and the information society: introduction. In Rayward, W. Boyd ed. *European modernism and the information society: informing the present, understanding the past*. Aldershot: Ashgate, pp. 1–26.

Rayward, W. Boyd. 2008b. The march of the modern and the reconstitution of the world's knowledge apparatus: H.G. Wells, encyclopedism and the world brain. In Rayward, W. Boyd ed. *European modernism and the information society: informing the present, understanding the past*. Aldershot: Ashgate, pp. 223–40.

Smiraglia, Richard P. 1990. New promise for the universal control of recorded knowledge. *Cataloging & classification quarterly* 11n3/4:1–15.

Smiraglia, Richard P. 2007. Two Kinds of Power: insight into the legacy of Patrick Wilson. In ed. Dalkir, Kimiz and Arsenault, Clément eds. *Information sharing in a fragmented world: crossing boundaries: Proceedings of the Canadian Association for Information Science annual conference May 12–15, 2007*. http://www.cais-acsi.ca/2007proceedings.htm.

Smiraglia, Richard P. 2013. Is FRBR a domain? Domain analysis applied to the literature of *The FRBR Family of Conceptual Models. Knowledge organization* 40: 273–82.

Van den Heuvel, Charles. 2008. Building society, constructing knowledge, weaving the web: Otlet's visualizations of a global information society and his concept of a universal civilization. In Rayward, W. Boyd ed. *European modernism and the information society: informing the present, understanding the past*. Aldershot: Ashgate, pp. 127–54.

Wilson, Patrick. 1968. *Two kinds of power: an essay in bibliographical control*. Berkeley: Univ. of California Press.

Chapter 6
Taxonomy

6.1 Taxonomy—Defining Concepts

At the most basic level a taxonomy is an ordered list of terms together with their definitions or other determinant characteristics. Taxonomy is a way of defining the component entities in a domain. Some taxonomies are very well known to the general public—such as the periodic table used in Chemistry, the names of animal or plant species, the names of heavenly bodies, etc.—while others are used only in highly specialized domains. Few taxonomies are permanent, although many are very long-lasting, but even more are used only briefly to help categorize research results. Beghtol (2003) has referred to these latter taxonomies as "naïve classifications;" Hjørland (1997, 47) called them "ad hoc" classifications.

As we will see, the form and content of any taxonomy is dependent on the epistemology of the domain for which it has been developed. Another way of looking at this is to consider taxonomy as an extension of research in a domain. Research is always an epistemological exercise engaging phenomena from different paradigmatic points of view. This is a major reason why taxonomies differ dramatically— point of view dictates the approach to hypothesis generation, which in turn dictates the approach to data analysis. Hjørland (1997, 47) says that "the relationships between concepts is an epistemological question" and "knowledge can be organized in different ways and with different levels of ambition," which range from the low-level ad hoc classification, through the mid-level pragmatic, to the high-level scientific classification.

About now you might notice that there is some confusion about whether taxonomy is a form of classification. In the generic sense, meaning the assignment of phenomena to specific categories, taxonomy is a form of classification. In the literal sense in the domain of knowledge organization, a classification is an ontology organized according to a symbolic schema. Classification in this sense requires meaningful symbolic notation; this will be the subject of our next chapter. For now let us consider taxonomy a highly specific sort of ontology, that arrives along with the definitions of the charac-

© Springer International Publishing Switzerland 2014
R.P. Smiraglia, *The Elements of Knowledge Organization*,
DOI 10.1007/978-3-319-09357-4_6

teristics of the phenomena involved, and that also includes certain kinds of relationships, such as genus-species, etc. In Dahlberg's (2006, 15) classification of knowledge organization literature, taxonomies occupy a grouping on special objects and subjects, as opposed to universal classifications (e.g., bibliographic classifications such as the *Dewey Decimal Classification* or the *Universal Decimal Classification*).

6.2   Kinds of Taxonomies

6.2.1   *Natural Sciences*

The Linnaean Taxonomy (Linnaeus 1760) is used to classify living things, ranking organisms in a system of hierarchies, and requiring the name of both genus and species to identify in each case the class and member represented. The common example is *Homo sapiens*, for humans, which includes the class name *Homo* meaning "man" and *sapiens* meaning "sentient," or "knowing." The system uses a hierarchy of kingdoms, subdivided by phyla, divisions, classes, orders, families, genera, and species. The main thing to remember about this immense taxonomy is that it represents the empirical observations of scientists gathered over centuries. New names are added only with the agreement of international scientific communities and only after sufficient peer review. When a term is added, the genealogy of the science behind the term also is recorded, so that future scientists might know how to verify or refute future discoveries. Groups of entities are referred to as *taxa*, from the Greek for "order."

Here, as an example, is a taxonomy of instantiation derived from Smiraglia (1992, 28):

Simultaneous derivations. Works that are published in two editions simultaneously, or nearly simultaneously, such as a British and a North American edition of the same work. Often such simultaneous derivations will exhibit slightly different inherent bibliographic characteristics.

Successive derivations. Works that are revised one or more times, and issued with statements such as "second, [third, etc.] edition," "new, revised edition," works that are issued successively with new authors, as well as works that are issued successively without statements identifying the derivation.

Translations, including those that also include the original text.

Amplifications, including only illustrated texts, musical settings, and criticisms, concordances and commentaries that include the original text.

Extractions, including abridgments, condensations, and excerpts.

Performances, including sound or visual (i.e., film or video) recordings.

You see here an ordered list of terms; the order is one of prioritization bibliographically, those closer to the top of the list maintain closer proximity to the original ideational content, those near the bottom of the list represent changed content. You see also the definitions. The terms are collectively exhaustive because there are no more kinds of instantiation. And the terms are mutually exclusive, although they could occur in sequence, as for example if an abridgment were subsequently translated. The point is, this is a taxonomy generated empirically from research.

6.2.2   Typology

Curiously enough, the term typology is used for the same sort of arrangement when the entities involved are called types instead of *taxa*. Typologies are used in anthropology, archaeology, linguistics, theology, and psychology. In most instances, typologies are less robust scientifically than taxonomies, which means a type is assigned based on empirical observation but always is subject to change given analysis from future observations. Here we see Hjørland's notion of epistemological level at work. These typologies are assigned as working tools in the identification of artifacts as empirical research progresses. In the natural sciences, taxonomies are created at the end of the process, once hypotheses have been tested sufficiently to rule out all alternative explanations.

6.2.3   Knowledge Management

In knowledge management the term taxonomy is used rather loosely and along the lines of Hjørland's "ad hoc." If, for a given organization, an ontology represents its entire knowledge base, a taxonomy represents the vocabulary of a narrower group—an office, a business function, etc. Often the term is applied to lists of terms with no apparent thought to empirical exhaustivity or exclusivity.

6.3   Usage in KO

In knowledge organization, taxonomies are non-universal classifications, which means they are essentially classifications of phenomena in broad domains. As we have seen, these domains might constitute whole sciences, or they might constitute the knowledge bases of smaller discourse communities. In either event, taxonomies represent at the topmost hierarchical level a set of mutually exclusive and collectively exhaustive categories, such that all phenomena fall into one of the hierarchies and only one. Typologies, on the other hand, are classifications of characteristics of phenomena, and these need not be mutually exclusive nor collectively exhaustive. This means that a taxonomy emerges when empirical science has demonstrated sufficient consensus about the phenomena in question, but typologies emerge (rather as Beghtol (2003) suggests when she calls them "naïve" classifications) as empirical observation progresses. An interesting case of the admixture of the two occurs in Smiraglia (2006), in which a taxonomy of derivation is displayed alongside a typology of instantiation, the better to demonstrate the ubiquity of the phenomenon among information objects. This collocation can be seen in Fig. 6.1.

But, by placing typology alongside taxonomy, it is possible also to see the precise rate of consensus that exists from empirical analysis about the phenomenon at that point in time.

Bibliographic Works	Artifacts--Metadata	Artifacts--Representations	Personal Papers
simultaneous editions	-finding aids	-field photos	Photocopies
successive editions	-field notes	-working images	Carbon copies
predecessors	-letters	-exhibition color images	Photos
amplifications	-conservation treatment notes	-digitized exhibition images	-postcard with photo
extractions	-register descriptions; object cards	-conservation photos	-digitized scan of postcard with photo
accompanying materials	-image order invoices	-archived photographic negatives	-reprint of photo
musical presentation	-museum database records	-archived photographic prints	-digitized scan of photo
notational transcription	-catalog card records	-archived photographic transparencies	
persistent works	-finding aids		
translations		-object reproductions	
adaptations		-drawings	
performances		-3D models	

Fig. 6.1 Instantiation taxonomy (*left column*) and comparative typologies (*right columns*) (Smiraglia 2006, 386)

Several papers demonstrate taxonomy as the fruit of knowledge organization research. Ménard and Dorey (2014) present a bilingual taxonomy for image indexing. Souza et al. (2012) propose an integrative framework for a taxonomy of knowledge organization systems. DiMarco (2008) developed a taxonomy of learning objects. In each case, empirical analysis was used to generate a taxonomy based on a consensus measure of terms present in the literature of a domain. These papers also represent diverse methodological approaches. Di Marco used two established taxonomies to analyze the contents of an applications domain. Souza Tudhope and Almeida used a meta-analysis of prior research on knowledge organization systems to suggest complex dimensions. Ménard and Dorey used experimental techniques to analyze the efficacy of an emergent taxonomy when applied by users.

6.4 Summary: On Epistemology of Taxonomy

Essentially taxonomy is a first step toward the creation of classification. Rooted in empirical observation, taxonomies supply defining characteristics and identify the sources of the definitive science from which the characteristics were observed. Taxonomies, like ontologies, arrange concepts in hierarchical orders. Unlike ontologies, taxonomies usually represent the knowledge bases of easily definable domains, rather than larger universal bodies of knowledge. Typologies are a sort of taxonomy, using less well-established and more pragmatic observations. Together, these tools fuel the progress of knowledge by displaying gaps where research is required to extend understanding. The essence of taxonomy is its epistemological basis, which also is the cause for the general confusion in the use of the term "taxonomy."

References

Beghtol, C. (2003). Classification for information retrieval and classification for knowledge discovery: Relationships between "professional" and "naïve" classifications. *Knowledge organization* 30: 64–73.

Dahlberg, Ingetraut. 2006. Knowledge organization: a new science? *Knowledge organization* 33: 11–19.

DiMarco, John. 2008 (forthcoming). Examining Bloom's Taxonomy and Peschl's Modes of Knowing for classification of learning objects on the PBS.org/teachersource Website. In Arsenault, Clément, and Tennis, Joseph, eds. *Culture and identity in knowledge organization: Proceedings of the 10th International ISKO Conference, Montréal, 5–8 August 2008*. Advances in knowledge organization, v. 11. Würzburg: Ergon Verlag.

Hjørland. Birger. 1997. *Information seeking and subject representation: an activity-theoretical approach to information science*. New directions in information management 34. Westport, Conn.: Greenwood Press.

Linnaeus, Carolus. 1760. *Systema naturae*. Halle: Curt.

Ménard, Elaine, and Jonathan Dorey. 2014. TIIARA: a new bilingual taxonomy for image indexing. *Knowledge organization* 41: 113–22.

Smiraglia, Richard P. 1992. Authority control and the extent of derivative bibliographic relationships. Ph.D. dissertation. University of Chicago.

Smiraglia, Richard P. 2006. Empiricism as the basis for metadata categorization: expanding the case for instantiation with archival documents. In Budin, G., Swertz, C. and Mitgutsch, K., eds., *Knowledge organization and the global learning society; Proceedings of the 9th ISKO International Conference, Vienna, July 4–7 2006*. Würzburg: Ergon-Verlag, pp. 383–88.

Souza, Renato Rocha, Douglas Tudhope and Mauricio Almeida. 2012. Towards a taxonomy of KOS: dimensions for classifying knowledge organization systems. *Knowledge organization* 39: 179–92.

Chapter 7
Classification: Bringing Order with Concepts

7.1 The Core of Knowledge Organization

Classification is the quintessential core of knowledge organization. Like encyclope-dism, classification is a response to the impetus to create, expose, or impose order on that which is known. But leaving behind the explanatory capacity of an encyclo-pedia or even the definitive aspects of taxonomy, classification relies on structure to reveal the relationships that govern an ontological reality. In classification all of the tools of knowledge organization come into play: the point of view provided by epis-temology is revealed in the interplay between facets of comprehended ontology, semantic power is provided by the concepts enumerated and yet often is unfettered by language through the use of symbolic notation thus yielding potentially the ulti-mate interoperability, and syntax is provided by the overall structure of the classifi-cation, and in particular by syndetic structure, which links the components.

Svenonius says that classifications bring like-things together according to their attributes (2000, 10). Soergel says that classification provides a logically coherent framework (1985, 5). Beghtol (2010, 1045) extends these traditional definitions to demonstrate ways in which classifications serve as "cultural artifacts that directly reflect the cultural concerns and contexts in which they are developed." Hjørland reminds us that classifications serve different purposes and thus can themselves be classified into at least three groups (1997, 46):

- *Ad hoc classification (or categorization)*
- *Pragmatic classification*
- *Scientific classification.*

The difference is the level of ambition brought to the scheme by its creators. Ad hoc classifications are those that seem just to happen in a very useful way, such as the way your spices are arranged in your kitchen. Pragmatic classifications are more ambitious because they are designed to facilitate work—activity of some sort with a purpose. Thus the classification that grocers use to arrange items on the

© Springer International Publishing Switzerland 2014
R.P. Smiraglia, *The Elements of Knowledge Organization*,
DOI 10.1007/978-3-319-09357-4_7

shelves not only serves to keep brands together and like items in the same aisle for easier restocking, but it also serves to facilitate your hunt for a specific item. Scientific classifications are those that arise from research, and thus they represent the highest level of ambition, which is to control and also facilitate the discovery of new knowledge. We will work along Hjørland's levels as we look at the kinds of classifications that are most discussed in the KO domain, beginning with everyday classification (including folksonomy), then moving to naïve classification, and then at the highest level we will look at scientific classifications, especially those that arise from the bibliographic world. Finally, we will review "classification theory" in search of the elusive concept-theoretic that is said to drive all of knowledge organization.

7.2   Everyday Classification

Classification is a near universal human phenomenon. When you say hello to a child and she says "Grandma," it is because she has recognized that you are her grandmother, and therefore not any other person. She has created a classification with at least two categories—grandmother and not-grandmother—and she has assigned you as a member of one category and therefore not a member of the other. It is simple cognition at one level, but it is also classification. Classification permeates human activity.

In a formal sense classification is acknowledged to have two uses for scholarship. First, scientists use classification to order their phenomena of study—often called taxonomy, this activity is essential for the advancement of knowledge. The second major use of classification is for the ordering of useful knowledge, and this can be seen as activity that crosses a broad spectrum of uses from social to scholarly to bibliographic. Dahlberg (2006) points out the common methods in use, which are the designation of objects of interest (knowledge elements), designation of the conceptual parameters for categories and relationships among them (knowledge units; i.e., this is the building of ontology), and the mapping of entities to the designated structure (knowledge systems). Your doctor checks a box on a diagnostic form, you find the tomatoes in the Italian foods section of your supermarket, you find mystery next to biography at your public library—all of these are examples of the use of classification for the useful ordering of knowledge.

Sometimes this is called "every day classification," and it is rife throughout human experience. Every human action involves decision-making, which by its nature produces categorization; everything from the most simplistic (e.g., inside/outside, daytime/nighttime, hot/cold, safe/dangerous, etc.), to the complex (e.g., fruit/nut/meat, or cheap/costly/expensive but worth it, for example) sets up essential classifications of what might otherwise be considered intuitive knowledge. Jacob (2001) reviewed several approaches to understanding the human processes that contribute to a "cognitive core" (p. 81) and suggested that classification needs to be analyzed "by studying its impact within the settings of everyday activity" (p. 96). In an earlier paper Jacob (1994) emphasized the importance of categories as "building blocks of cognition" (p. 101). She also made an important distinction between the concepts of "categorization" and "classification," by contrasting the cognitive function of the former with the formal systematization of the latter. Humans engage in categorization constantly as part of the experience of being, but categorization itself is not classification. Classification provides structure according to a deliberate epistemology. A system of formal constraints are imposed on the categories, and these constraints embody cultural assumptions (or epistemological demands).

7.3 Naïve Classification

The empirical derivation of knowledge-elements, particularly in developing or evolving KO systems, provides a basis upon which conceptual systems can be built. This is exactly Beghtol's process of naïve classification for use in evolving scholarship. According to Beghtol (2003, 66), the process of naïve classification has many uses, including the discovery of gaps in knowledge, the reconstruction of historical evidence, and the revision or amplification of existing knowledge organization schema, among others. The process requires the scholar to articulate the purpose for his work so as to limit the empirical parameters, and then a variety of techniques may be employed, including paradigm-identification, and ordering (hierarchy, tree-structure, faceting) techniques.

Beghtol refers to studies that report naïve classification of Chinese plates, paintings, religions, photographs, thirteenth century Spanish silks, and child-rearing practices, among others. Green and Fallgren (2007) use the techniques to analyze document structures for the revision of the *Dewey Decimal Classification*. Let us look at a very simple example. We begin by identifying the phenomena observed and creating simple groupings. Figure 7.1 shows a set of observations divided into two clusters.

In this illustration we have a naïve classification. There are nine objects in our laboratory, and they clearly can be divided by observable likeness into two categories, hearts and suns. There are five suns and four hearts. This classification is complete, because it identifies the knowledge elements (objects of study), knowledge units

suns   **hearts**

Fig. 7.1 Suns and hearts (Smiraglia 2009, 9)

(hearts and suns), and a knowledge system "Suns and Hearts." The classification is naïve because it represents merely the grouping of observations in this particular instance. We do not yet know why the suns are not hearts or why the hearts are not suns, we do not know what the suns and hearts have in common except that they both are found in this observation, we do not know why there are more suns than hearts or fewer hearts than suns, and we do not know why there are not other objects, such as moons or lightening bolts or stars. In research, the use of naïve classification is intended for just this process—to identify the paradigm and to create a matrix of its contents from which hypotheses can be generated to structure future research.

7.4 Classification Systems

Classification systems are devised for many purposes, not the least of which is the imposition of order on a domain of activity or productivity in society. Elichirigoity and Malone (2005) detail the evolution of the North American Industry Classification System, which was designed to accompany the shift from an industrial to a service economy. Not surprisingly, the epistemological impact of the classification lies in its influence on the enforcement of a specific conception of economic production and value. Bowker and Star (1999) similarly describe the evolution of classifications with social consequences, ranging from the classification of race that accompanied South African apartheid, to the history of the International Classification of Diseases (from which emerged the absurd historical fact that in the nineteenth century more people died of apoplexy than cancer), to the Nursing Intervention Classification, which forces nurses to describe their work according to a predetermined schema rather than according to the actual care provided. Classifications exist to organize knowledge, but also to influence its use.

The same is true of bibliographic classifications. Just as those devised by philosophers (see Chap. 2) such as Bacon or Foucault are designed to influence the comprehension of knowledge, bibliographic classifications are designed to influence the organization of recorded knowledge. Although professions such as librarianship rely on bibliographic classifications as primary pathways to information retrieval, the fact is that the epistemological assumptions that underlie their structures do as much to inhibit resource discovery as to influence it.

Melville Dewey's magnificent world-reknowned classification comes under significant criticism because of the way in which it forces knowledge to be organized according to the cultural norms of white, male, western society. Olson (1998) demonstrates the denigration of women, Puerto Ricans, Chinese-, Japanese-, and Mexican-Americans, Jews, Native Americans, the entire developing world, gays, teenagers, seniors, people with disabilities—none of them fare well in Dewey's classification. Why? Because their marginalization is reflected in the literature that is collected deliberately by information institutions, such as the Library of Congress, who hold authority over the dissemination of knowledge throughout American culture. This is reflected in the *Dewey Decimal Classification* (*DDC*), because the rule of literary warrant insists that only knowledge held in books collected by libraries may be included, and it must be included in a way that reflects the opinion of the authors of the books involved. Does this sound like a circular argument? Yes, of course. But so does most social discrimination—"you cannot be equal, because heretofore you never have been"—or "we've always done it this way."

Furner (2007) used critical race theory to demonstrate how *DDC* could be deracialized, offering for the first time a workable solution to a more egalitarian classification. Furner says (2007, 165):

> We might consider that any decision taken to prevent classifiers and searchers from the use of racial categories is to ignore an everyday reality in which those categories are invoked not only in the distribution of social and political power, but also in individuals' self-identification.

Although, bias has its cultural role, as Hjørland points out (2008) in describing the cultural influence of the placement of concepts in classification. His best example? The Canary Islands briefly belonged to Denmark. In Danish libraries, they are classified as part of Denmark. Is that bias? Or is it cultural collocation for Danish library users?

7.5 Properties of Classifications

Classifications must be inclusive as well as comprehensive, which means a given classification must include all possible entities within its field of coverage. A simple example might be a classification of pets. It should have not only cats and dogs but also Siamese cats and hounds. A classification must encompass all collectible resources within its field of interest. Like controlled-vocabularies, classifications are expected to employ terminology that is clear and descriptive with meaning that is consistent for both the user and the classifier.

Classifications must be systematic, which means there must be rules of inclusion and exclusion that are easily understandable as well as applicable. Classifications are also supposed to be flexible and expansible, which means that as new entities are discovered there must be space for them and rules that allow them to be incorporated. This means that classifications, like their cousins controlled-vocabularies, are sensitive to cultural changes in point of view as well as to new discoveries, so they are constantly being updated.

Enumerative classifications attempt to assign designations for all single and composite subject concepts required in the system. Every concept that must be represented must have a location in the classification. Hierarchical classifications are those that are arranged according to the principle of general-specific relations. For example, Fig. 7.2 shows a simple hierarchy of banking (we called this a domain-specific ontology in Chap. 5)

This is a hierarchy proceeding from general at the top to specific at the bottom. Classifications differ from ontologies in one important way—they are arranged according to symbolic notation. Notation allows the ontology to retain its logical

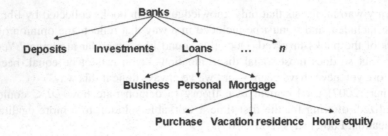

Fig. 7.2 Hierarchy, or ontology, of banking

ordering of concepts regardless of the semantics of natural language. For this reason, classifications also often come with alphabetical indexes. Notation might be expressive, meaning it functions like a language (for example, in *DDC* "92" always means history or biography), or it might simply be logical. For example, if we give each node in our banking ontology a number we have the schedule (its map) for a notated classification:

1 Banks
1.1 Deposits
1.2 Investments
1.3 Loans
1.3.1 Business
1.3.2 Personal
1.3.3 Mortgage
1.3.3.1 Purchase
1.3.3.2 Vacation residence
1.3.3.3 Home equity
1.*n other nodes as necessary*

This classification is hierarchical, proceeding from classes to divisions to subdivisions, and following a logic of subdivision. It also is expansive, because of its decimal structure any new concept can be entered in future as necessary.

Synthetic is a term used to mean a certain kind of flexibility in which different parts of a classification may be used together to express complex subjects. Imagine we also had a classification for houses, in which the term 06 Townhouse existed. If our classification were synthetic we could then express the concept of "Mortgage for a townhouse" by adding together the terms 1.3.3 for mortgage and 06 for townhouse to get 1.3.3-06. Synthetic classifications assign designations to single, unsubdivided concepts and give the classifier generalized rules for combining these designations for composite subjects.

Classifications also may be faceted to allow the combination of several different classification symbols in a prescribed sequence, in order to express clearly defined, mutually exclusive, and collectively exhaustive properties, or characteristics of a subject. The Universal Decimal Classification and the Bliss Bibliographic Classification are two examples of universal faceted classifications that allow almost any combination of concepts to be expressed. Marchese and Smiraglia (2013, 256) use the following example (abbreviated here) to demonstrate the flexibility of the UDC's faceted structure.

In UDC the symbol 625.714 means "towpaths." It falls within a hierarchy:

6 Applied sciences
62 Engineering
625 ...
625.71 Kinds of ordinary road according to importance and purpose
625.714 Roads along watersides (embankments). Causeways. Towpaths

UDC facets are added using the connecting symbols "+" or "/" or ":" to add dimension to a conceptual representation adding symbols from so-called auxiliaries or even by adding concepts together. So a towpath in New Hope, Pennsylvania might add 734.811.4 Bucks County thus:

625.714(734.811.4) Towpaths in Bucks County, Pennsylvania, US.

This, however, does not tell us whether it is a towpath in 2013 with tourists sitting along it, or a towpath in 1864 with donkeys pulling armaments for the American Civil War. We could add a dimension of time thus:

625.714(734.811.4)"1864"

and now we have expressed a place and a time, but still not whether we are dealing with building a towpath (the implication of 626.32 Hydraulic engineering) as opposed to 625.714 for kinds of roads, or whether we mean instead navigating a towpath. We could add 536.78 "Journey in straight line" to show we mean navigating:

536.78 + 625.714(734.811.4)"1864"

Another structural theory of classification is called the theory of integrative levels. This notion replaces hierarchy with an evolutionary progression from the simple to the complex according to the accumulation of properties (Beghtol 2010, 1055).

7.6 Concepts Well in Order

In sum it is easy to see even from these simplistic examples how complex classifications emerge. The assignment of observations to categories is a basic human intellectual function, which extended to its logical use in knowledge organization leads to the development of major systems for ordering. These systems are culturally ubiquitous, and therefore it is critical to understand how they emerge, evolve, and grow into both useful systems for information storage and retrieval and oppressive agents of bias. Concept-theoretic (Dahlberg 2006) is the basis for the ontology generation that is the beginning of all classifications. Yet, as we have seen in earlier chapters, there can be no single appropriate set of concepts, because all understanding is perceptual. The best we can do is to comprehend the innate orders of concepts in every domain, the better to seek pathways for interoperable understanding.

References

Beghtol, Clare. 2003. Classification for information retrieval and classification for knowledge discovery: Relationships between "professional" and "naïve" classifications. *Knowledge organization* 30: 64–73.

Beghtol, Clare. 2010. Classification theory. In Marcia J. Bates and Mary Niles Maack eds., *Encyclopedia of library and information sciences*, 3rd ed. Boca Raton, FL: CRC Press 1: 1045–60.

Bowker, Geoffrey C., and Susan Leigh Star. 1999. Sorting things out: classification and its consequences. Cambridge: MIT Press.

Dahlberg, Ingetraut. 2006. Knowledge organization: a new science? *Knowledge organization* 33: 11–19.

Elichirigoity, Fernando, and Cheryl Knott Malone. 2005. Measuring the new economy: industrial classification and open source software production. *Knowledge organization* 32: 117–127.

Furner, Jonathan. 2007. Dewey deracialized: a critical race-theoretic perspective. *Knowledge organization* 34: 144–68.

Hjørland, Birger. 2008. Deliberate bias in knowledge organization? In Arsenault, Clément, and Joseph T. Tennis, eds., *Culture and identity in knowledge organization: Proceedings of the Tenth International ISKO Conference 5–8 August 2008 Montréal, Canada*. Würzburg: Ergon-Verlag, pp. 254–61.

Green, Rebecca, and Nancy Fallgren. 2007. Anticipating new media: a faceted classification of material types. In Tennis, J. ed. *North American Symposium on Knowledge Organization* http://dlist.sir.arizon.edu/1911.

Hjørland, Birger. 1997. *Information seeking and subject representation: an activity-theoretical approach to information science*. New directions in information management 34. Westport, Conn.: Greenwood Press.

Jacob, Elin K. 1994. Classification and crossdisciplinary communications: breaching boundaries imposed by classificatory structure. In Albrechtsen, Hanne and Oernager, Susannne eds. *Knowledge organization and quality management: Proceedings of the Third International ISKO conference, 20–24 June, 1994, Copenhagen, Denmark*. Advances in knowledge organization 4. Würzburg: Ergon, pp. 101–8.

Jacob, Elin K. 2001. The everyday world of work: two approaches to the investigation of classification in context. *Journal of documentation* 57: 76–99.

Marchese, Christine, and Richard P. Smiraglia. 2013. Boundary objects: CWA, an HR firm, and emergent vocabulary. *Knowledge organization* 40: 254–59.

Olson, Hope A. 1998. Mapping Beyond Dewey's Boundaries: Constructing Classificatory Space for Marginalized Knowledge Domains. *Library trends* 47 no. 2: 233–55.

Smiraglia, Richard P. 2009. Defining bibliographic 'works': naïve classification for terminology generation. In Catalina Naumis Peña, ed. *Memoria del I Simposio Internacional sobre Organización del Conocimiento: Bibliotecología y Terminología*. México, D.F.: Universidad Nacional Autónoma de México, pp. 7–17.

Soergel, Dagobert. 1985. *Organizing information: principles of data base and retrieval systems*. Orlando: Academic Press.

Svenonius, Elaine. 2000. *The intellectual foundation of information organization*. Cambridge, Mass.: MIT Press.

Chapter 8
Metadata

8.1 The Roles of Metadata

Metadata are descriptive terms that are applied to information resources, primarily
for the purpose of facilitating retrieval. If I say this book is red, and you ask the
system for a red book, a match will occur and everybody is happy. Were that the
problem were actually so simple. In fact, metadata are used in a variety of ways in
resource description and thus potentially play different roles in knowledge organization. Let us begin with a simple example, a citation for a monograph, formulated
according to the *Chicago Manual of Style*:

Smiraglia, Richard P. 2001. *The nature of a 'work:' implications for the organization
of knowledge*. Lanham, Md.: Scarecrow.

In this simple format metadata serve as descriptors of the book (the physical
item) and also of the work by Smiraglia printed in the book. Traditional metadata
for citing sources in publications are author name, date of publication, title, subtitle,
place, and publisher. These data are considered sufficient to recognize the book
when its citation is located in an information retrieval system. They also are considered sufficient for acquiring the book, say by placing an order for it or by looking
for it at an online bookseller site.

The data are considered unique as well. For instance, there are few people named
Smiraglia, and even fewer named Richard P. Smiraglia, and fewer still who wrote a
book in 2001, and only one who wrote a book with this title. So the author identifier
is sufficiently discrete to allow for high precision both in assigning the name and in
searching for it. The same logic would apply to the title of the book, especially in
context with its subtitle. The place and publisher are not unique to this item, of
course, and the date is not unique at all, but contextually speaking they provide
discrete data. This is resource description at its simplest, in the form in which scholars routinely practice it by referencing source material in their writing. It is however,
by itself, not knowledge organization.

© Springer International Publishing Switzerland 2014   65
R.P. Smiraglia, *The Elements of Knowledge Organization*,
DOI 10.1007/978-3-319-09357-4_8

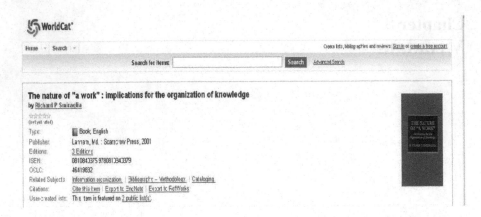

Fig. 8.1 WorldCat Metadata

Metadata for resource description are considered to play a role in knowledge organization when they are used to provide order to a set of such descriptions. If we place this citation among several others, then we will have to choose an entry element. We might enter it, as above, under the author's surname. If we do, we have collocated this with other writings by the same author. We also have created a set of writings by that author, which will be distinct from writings by other authors, and which likely will be subarranged within the set by date or title. This creates what has been called an alphabetico-classed arrangement, in which a class of materials is identified (works by one author) so as to collocate as well as to disambiguate by separating them from works by other authors, and, divisions within the class are created to keep order among the person's works by title, or by chronology, or both.

The same is true of library catalogs as well, but the metadata descriptions tend to be more complex. Here is a description of the same book taken from the OCLC WorldCat (Fig. 8.1):

The dataset is similar but has more detail. For instance, there is an ISBN (International Standard Book Number) which is useful for ordering the book but also for controlling metadata sets (it is much easier to search a system for a unique number than for any combination of terms). There also is an OCLC number, which is the internal number of the bibliographic record that represents this book in the WorldCat. It is just an inventory device, but also very helpful for controlling sets of bibliographic metadata. Also you see something called "related subjects." These actually are subject headings applied from the *Library of Congress Subject Headings*; these are subject descriptors from a highly controlled, pre-coordinated vocabulary. A form of metadata themselves, in this case they are used to group bibliographic records for books together with other books that are similar in topical treatment. You might also notice a photograph of the book, which is relatively new in library applications, but is another form of identifying metadata.

The same thing can be accomplished, of course, by adding subject descriptors to simple citations. In indexing services (databases such as LISA, for example) citations are accompanied by terms from thesauri, and sometimes just by keywords from an abstract supplied by the author or an indexer. These are post-coordinated vocabulary terms, meaning it is up to the searcher to combine sufficiently relevant terms in the search process to yield a fruitful result. There also are terms added by system participants, known as bookmarks or tags or social tags; these are common in Web 2.0 applications. Here are some tags for this book found in Amazon.com:

Books about books (12)
Library science (7)
Gifted (6)
Information retrieval (6)
Librarian (6)
Library research (6)
Itil v3 (5)
Memoir (5)
School library (5)
Bibliophile (3)

Presumably these terms have meaning to the people who assigned them, although a couple of them seem a bit obscure. These terms can be used for subject searching, like keywords, but they also provide a somewhat more social search capability, which is why they sometimes are referred to as folksonomy. Term assignment has few or no rules at all, which means terms are collectively less discrete than those assigned using pre- or post-coordinated controlled vocabularies. Ultimately there is a trade off between the guerilla freedom associated with social tagging and the lack of precision. At any rate, all subject descriptors are considered to be a kind of knowledge organization, because they all can be used to identify the conceptual aspects of recorded knowledge, and then to group knowledge according to the concepts identified.

So far we have seen metadata play two roles: as identifiers for resources that are artifacts of recorded knowledge, and as ordering devices either for those artifacts or for the conceptual material they convey. There is one more form of metadata commonly used in knowledge organization, and these are metadata schemas that provide carrier shells for collected bibliographic data. That is, some metadata are used to format other metadata for the purpose of populating retrieval systems. We can look at two examples. First, Fig. 8.2 shows the MARC (Machine-Readable Cataloging) record for our book.

A MARC record contains at least three discrete data sets. First, it contains the legible metadata about the book that appeared in the WorldCat record above. Descriptive terms are derived from the item itself according to the *Anglo-American Cataloguing Rules*, and are supplemented with subject descriptors from *Library of Congress Subject Headings*, and with bibliographic classification numbers from both the *Dewey Decimal Classification* and the Library of Congress *Classification*. These data are formatted using MARC conventions that serve to facilitate machine-manipulation of the data. MARC fields are structured using three-digit field-tags, one or two digit field level indicators, and subfield indicators that delimit the field. Field 245 identifies the title and statement of responsibility of an item, opens with

| OCLC | 46419832 | | No holdings in ZPS - 206 other holdings; 7 other IRs | | | | |

| Books ▼ | Rec Stat c | Entered 20010207 | Replaced 20080401175245.8 |

Type	a	ELvl		Srce		Audn		Ctrl		Lang	eng
BLvl	m	Form		Conf	0	Biog		MRec		Ctry	mdu
		Cont	b	GPub		LitF	0	Indx	1		
Desc	a	Ills	a	Fest	0	DtSt	s	Dates	2001		

>010 2001020328
>040 DLC $c DLC $d UKM $d C#P $d BAKER $d BTCTA $d YDXCP $d OCLCG
>015 GBA1-W8265
>019 50175847 $a 51000269
>020 0810840375 (alk. paper)
>020 9780810840379 (alk. paper)
>050 00 Z666.5 $b .S47 2001
>082 00 025.3 $2 21
>090 $b
>049 ZPSA
>100 1 Smiraglia, Richard P., $d 1952-
>245 14 The nature of "a work" : $b implications for the organization of knowledge / $c Richard P. Smiraglia.
>260 Lanham, Md. : $b Scarecrow Press, $c 2001.
>300 xviii, 182 p. : $b ill. ; $c 22 cm.
>504 Includes bibliographical references (p. 135-143) and index.
>650 0 Information organization.
>650 0 Bibliography $x Methodology.
>650 0 Cataloging.
>938 Baker and Taylor $b BTCP $n 2001020328
>938 Baker & Taylor $b BKTY $c 58.50 $d 58.50 $i 0810840375 $n 0003700931 $s active
>938 YBP Library Services $b YANK $n 1755428
>029 1 UKM $b bA1W8265
>029 1 UKM $b bA246498
>029 1 YDXCP $b 1755428
>029 1 NZ1 $b 6149340
>029 1 AU@ $b 000022503905

Fig. 8.2 MARC Metadata Set

two indicators that help the system to index the particular title, and (in this case) has two subfield indicators designated with delimiter signs represented here as $-signs. The *AACR* description runs from field 100 through field 504; the *LCSH* are in fields numbered 650, the classification numbers are in fields 082 and 090 respectively. The remaining fields are called "control" fields because they contain ISBN and OCLC record numbers, and other identifiers useful for machine control of the dataset. The MARC schema is a wrapper, or a data format schema, that is used as a carrier for bibliographic data composed using other schema. This is a very complex process, and it reflects centuries of evolution in the structure of library catalogs.

On the other hand, a simpler metadata schema has been developed, called "Dublin Core," that is intended to provide better web-based metadata representation. Figure 8.3 is the DC representation of our book.

The Dublin Core schema is intended to reside in web-based indexing locales, and is a very simple schema. In this case the same data from the MARC record are represented and identified using DC "name" conventions inside the DC wrappers. Are MARC or Dublin Core (or RDF or other metadata schemas) forms of knowledge organization? When they are used as the backdrop for conceptual ordering

```
<meta name="DC.Title" content="nature of "a work" : implications for the organization of knowledge /">
<meta name="DC.Creator.namePersonal" scheme="MEntry" content="Smiraglia, Richard P., 1952-">
<meta name="DC.Format.extent" content="xviii, 182 p. : ill. ; 22 cm.">
<meta name="DC.Publisher.place" content="Lanham, Md. :">
<meta name="DC.Publisher" content="Scarecrow Press,">
<meta name="DC.Date.issued" scheme="MARC21-Date" content="2001">
<meta name="DC.Identifier" scheme="LCCN" content=" 2001020328">
<meta name="DC.Identifier" scheme="ISBN" content="0810840375 (alk. paper)">
<meta name="DC.Identifier" scheme="ISBN" content="9780810840379 (alk. paper)">
<meta name="DC.Language" scheme="ISO639-2" content="eng">
<meta name="DC.Subject.class" scheme="LCC" content="Z666.5 .S47 2001">
<meta name="DC.Subject.class" scheme="DDC" content="025.3">
<meta name="DC.Subject.topical" scheme="LCSH" content="Information organization.">
<meta name="DC.Subject.topical" scheme="LCSH" content="Bibliography--Methodology.">
<meta name="DC.Subject.topical" scheme="LCSH" content="Cataloging.">
<meta name="DC.Type" scheme="OCLCg" content="Text data">
```

Fig. 8.3 A Dublin Core Metadata Set

of bibliographic records they are clearly a form of knowledge organization. And because they are ubiquitous in information retrieval, their structure affects the efficacy of other forms of conceptual ordering. Many scholars have looked at problems of metadata generation (for example Greenberg 2005, and Smiraglia 2006), and many others (most notably Howarth; see for example 2002 and 2006) have investigated the problems of using metadata to address problems of semantic interoperability on the web.

We will consider metadata for subject (or conceptual) ordering in later chapters of this book. In this chapter we will look at the problems surrounding these other roles for metadata in knowledge organization. In terms of resource description, we will look at developments leading to the reprioritizing of bibliographic entities with works as the more important member of the works-items pair. We will look also at ways in which metadata schema are now serving as platforms for semantic interoperability. And we will also consider the epistemological ramifications in the derivation of metadata automatically, pragmatically, and rationally.

8.1.1   What Is a Text?

The fundamental phenomena in information, and therefore in knowledge organization operating as a subdomain within information, are documents and their contents. Every tool, even at the most philosophical metalevel, is constructed with the objective in mind of pointing people to informative objects, usually called documents. Hjørland (2003, 89) says "LIS is not primarily focused on constructing algorithms, but on informing people about documents." Therefore a foundational question for knowledge organization is "what is a document?" Buckland (1997) has suggested that documents are socially constructed sources of evidence; Hjørland (1997) has similarly suggested that documents are defined not by their physical form but by the use to which humans put them. So we see that regardless of the physical form in which we might find a document, the term itself is a generic term for a specific kind of recorded knowledge. Documents are generally considered to be evidentiary, and thus are epistemologically more closely allied with data than with, say, synthesized texts.

So the next question, then, is what is a text? A text is a set of semantic strings that communicate ideational content. When played back, either by being read (in which case they are "played" in the reader's consciousness) or by being reproduced auditorially, the strings of a text exist to record the evidence that constitutes a document. Beyond this, there is much confusion about what, exactly, constitutes a text Smiraglia (2001a, 3), defines a text as "the set of words that constitute a writing. A text is not the same as a document, which is the physical container (an item) on which the text is recorded. A document may have only one text, but a text may appear on many documents." Text, then, is another generic term that denotes the communicative aspect of the evidentiary value of a document.

Texts may be texts of works (about which more shortly), or they may simply be texts of uncoordinated writings. In either case, texts are realized during playback, and in this process are relatively unstable, despite the stability of their physical form. Barthes (1975) has famously referred to text as "tissue," meant to be broken, torn, and shoved away. People, he says, skim or skip, boldly or accidentally, noticing not so much definitive content as the "abrasions" on the surface. Finally he says (p. 64) that the essence of a text is the product of the interweaving of its tissue—like a spider's web, perpetually rewoven in different form as the product of experience.

8.1.2   Then What Is a Work?

Smiraglia (2001a, 95) says a work is:

> The set of ideas created by an author or other artist, set into a document using text, with the intention of being communicated to a receiver (probably a reader or a listener ...). A work may have many texts, and may appear in many documents and even in many documentary forms.

In the long history of resource description for the development of library catalogs, the "work" has always played a central role, yet before this the concept was never adequately defined, which led to quite a lot of confusion. If we recall Cutter's "objects," it is easy to note that the focus of his system was a "book." Explicit is the notion that a book contains a work. But in reality, the bibliographic universe is not so simply constructed, nor has it ever been. Even in Callimichus' day, more than one work might have been recorded in a single book. The same has been true all through bibliographical history. We need look no farther than the testimony of Panizzi, who in 1848 said that no catalogue could be considered useful unless it allowed a choice among editions of a work (Panizzi [1848] 1985, 21). That the primacy of works as entities of choice could be the subject of legislative hearings is just one example of the degree to which controversy has surrounded this phenomenon. A century later with regard to development of post-World War II cataloging rules Lubetsky said essentially the same thing, that searchers who hunt for books are more likely to be interested in the works that are contained in them (Lubetzky 1969, 11). And yet again, in 1987, Wilson pointed to the work as the subject of Cutter's second objective and not his first.

The evidence in the studies Smiraglia reports demonstrates the degree to which that entity called a "work" might, in fact, exist in multiple instantiations produced over time according to a variety of cultural catalysts. In a paper about the epistemology of the work, Smiraglia (2001b, 192) expands the definition to embrace these cultural concepts:

> Works (e.g., musical works, literary works, works of art, etc.) are key entities in the universe of recorded knowledge. Most recorded knowledge survives through authored (or otherwise created) entities. Works are those deliberate creations (known variously as opera, oeuvres, Werke, etc.) that constitute individual sets of created conceptions that stand as the formal records of knowledge.

An operational definition of a work became necessary for empirical research to quantify the extent of membership of texts in sets of instantiations of works. Works have semiotic character—that is, they have iconic character by virtue of the collective consciousness about them, such that they can be seen as signifying entities. This means, if a work becomes known among a group of people their common perception of it can take on a life of its own, quite apart from any concrete textual content. For example, if you say you have seen *Gone with the Wind* you mean that you have seen the motion picture made from the screenplay derived from the novel by Margaret Mitchell. It likely was the visual impact of the motion picture that has impressed you, and this takes on a life of its own among those who discuss it and keep the idea of it alive in public consciousness. The concept of *Gone with the Wind*, then, is the sign (or icon, to use Peirce's term) associated with the work. The icon is used to evoke or trigger associations that are linked to the work. The work itself, the motion picture, is part of a larger family of works, all having closely-related ideational content (that is, sets of ideas), all called *Gone with the Wind* and all derived directly or otherwise from Margaret Mitchell's novel. You might never have read the novel, you might never have seen the movie, but you still likely will know what it is because of this signifying characteristic.

Empirically speaking, once a work enters the public consciousness it becomes part of a canon, and that canonization seems to serve as the catalyst for the development of large numbers of instantiations. Smiraglia (2007a) reports on this phenomenon observed in a study of best-selling books. Surprisingly, not all "best-sellers" generate large instantiation sets. Rather, only those that enter the cultural canon do so. Although all best-sellers appear in more than one edition initially, which seems to be part of the publishing strategy to make a best-seller, only a small proportion last beyond the first year of popularity and go on to become works with many translations or editions, or are reworked as screenplays from which motion pictures are made, etc. These large sets of multiply instantiating works have been called "super-works" by Svenonius (2000). Smiraglia (2007b) explains how they grow with examples as diverse as a Shostakovich string quartet and Annie Proulx's short story *Brokeback Mountain*. The generic phenomenon of instantiation—meaning the

Fig. 8.4 Instantiations
Arranged by Uniform Title

Dickens, Charles, 1812-1870.

[Oliver Twist]
[Oliver Twist. Chinese]
[Oliver Twist. Dutch]
[Oliver Twist. French]
[Oliver Twist. German]
[Oliver Twist. Hebrew]
[Oliver Twist. Japanese]
[Oliver Twist. Spanish]
[Oliver Twist. Ukrainian]

...

Bart, Lionel.

[Oliver! Selections]

coexistence of multiple realizations of a work over time—is demonstrated to apply not only to works but also to other kinds of information objects (such as museum artifacts and archival documents) in Smiraglia (2008). The instantiations, themselves usually discrete publications, have been classified by Smiraglia (2001b) as either mutations or derivations. An instantiation is called a derivation when it clearly shares most of the ideational *and* semantic content with its predecessor, as is the case with many successive editions, or abridgments, for instance. An instantiation is called a mutation when it is only distantly related though ideational content and its semantic content is mostly new, as is the case with translations, screenplays, adaptations, and so forth. In Smiragli (2006) these categories are shown to constitute a typology of instantiation at a meta-level, such that the characteristics are not mutually exclusive but can occur together in a single instantiation. For example, an adaptation of a novel for a juvenile audience would be a mutation, but its second edition would be a derivation of a mutation.

The problem for metadata, clearly, is two-fold: (1) to collocate like instantiations and relate them to their progenitor works; and (2) to disambiguate large retrieval sets by making clear the distinctions among instantiations. The following example (Fig. 8.4) appears in Smiraglia (2001b), to demonstrate the manner in which current catalogs accomplish this task.

Here you see headings (collocating devices) that consist of author names and uniform titles. Each presumably identifies one or more physical publication. All of them have the title keyword "Oliver" and most of them "Oliver Twist." Without the artificial headings collocation likely would not take place. And without the subheadings for language, the translations would be all mixed together without differentiation from the English-language editions.

Since 1998 the library community has embraced an entity-relationship framework for reimagining and reconstructing the library catalog. Called *Functional Requirements for Bibliographic Records* (IFLA 1998) or *FRBR* for short, this new framework places the work as the primary objective of information retrieval, with

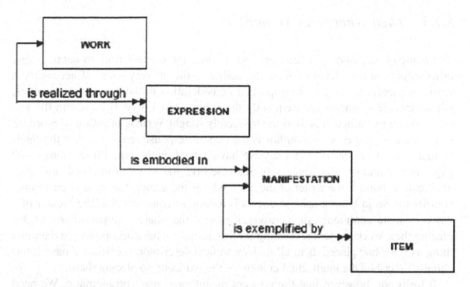

Fig. 8.5 FRBR Entities and Relationships (IFLA 1998, 13)

its physical instantiations divided according to their role in realization, whether intellectual or physical. *FRBR* itself (1998, 12) refers to these entities as representative of "user interests in the products of intellectual or artistic endeavour." The entities are work, expression, manifestation, and item, as can be seen in the representation reproduced in Fig. 8.5.

By reading the diagram from top to bottom one can grasp the rationalized set of relationships represented. A "work" is an intellectual creation that can be realized when it is expressed. This expression can be embodied in a manifestation, which can be exemplified by a specific item. Smiraglia's *The nature of a work* is a work at the level at which he conceived it, and its text is its expression of ideas in semantic form, which becomes manifest when gathered together under this title, and can then be exemplified by copies of the red book we saw above. The scheme is derived rationally, based on prior cataloging rules, and does not accord with the empirical evidence reported by Smiraglia and others in the studies referred to above (the empirical evidence suggests there are only three levels, and that the middle level defines a domain described by an as of yet incomplete typology including at least mutations and derivations). Nevertheless it represents a step forward for information retrieval, by separating inventory control of physical items from intellectual control of works. One can envision a search engine that allows exploration among works and their expressions, and then selection of a specific item for physical use.

The problem for metadata as an aspect of knowledge organization is how to organize works and their instantiations apart from the problems of inventory control of specific items. The traditional approach in library catalogs has been to assign works to a heading representing their creators (authors mostly) and then to subarrange in various ways the editions that are so gathered. All of this brings us to an interesting related question—what is an author?

8.1.3   Then What Is an Author?

The name of an author is a "concept" for knowledge organization. In catalogs and other information retrieval devices, the author is the primary route of access for a work. Arguably this is a more important consideration in the arts and humanities, where specific syntheses are frequently the subject of a query. But even in the sciences, works by particular authors are frequently sought. What is an author? According to most cataloging codes, an author is the person responsible for creating the intellectual content of a work. Let's say we have a book with some illustrations—400 pages of text and six lithographs—in this case both the writer of the book and artist are creators, but it is the writer of the book who is the author; the artist is an adjunct contributor. So in bibliographical circles is an author constructed as the creator of a work, or more commonly, of a corpus of works, that share a certain character. By placing these works under the heading for this author we have accomplished the same thing as if we had placed them all under a topical descriptor, because we have inferentially described the intellectual content of the works by so placing them.

It turns out, however, that the concept of author is itself problematic. We need only look at Foucault (1984) who suggests that the entire concept is a social construct. This makes quite a lot of sense after all, because none of us knows Ernest Hemingway, but quite a few of us know the person who seems to be the soul of the works that are said to have been written by Ernest Hemingway. It reminds us of the old argument about who wrote Shakespeare. Well, Shakespeare did, of course, whether it was the person named that or not—the concept of Shakespeare as an author is a social construct. And therefore, it functions in knowledge organization in the manner of a concept. Foucault (1984, 101) says that an author is the figure to whom a given text "points." A writing, in Foucault's notion (1984, 102), is an "interplay of signs arranged less according to its signified content than according to the very nature of the signifier. Writing unfolds like a game." He means, rules of inclusion and exclusion apply to the attribution of a work, and primary among them is the iconic nature of an author's work. That is, a work is not the text you are reading, so much as it is what that text represents. Therefore, an author is also not the person who writes a text, but rather the token in the game of creating the iconic work. Hemingway's *A Moveable Feast* is so much more literarily palatable than, say, Martha Sterchongish's. Filiation with Hemingway is the prize in this game; if we say we know Hemingway and our stated knowledge is socially accepted then we have "won" the game for the moment and mounted the winner's circle. Finally (p. 108):

> The author function is therefore characteristic of the mode of existence, circulation, and functioning of certain discourses within a society.

An author is a social attribution, and not a real individual, but simply a cultural sign, a player in the game of discourse.

In library cataloging, attribution of works to authors has a tradition stretching back to antiquity. But only in the late nineteenth century did the "authorship principle" gain its name. Julia Pettee was its staunchest defender. She suggested the role of an author was to create what she called a "literary unit." A literary unit is what we now call a work, but Pettee also intended the term in its functional sense as we understand it in knowledge organization, as a gathering point for conceptual organization. A recent series of papers by Smiraglia, Lee and Olson (2010, 2011) and Smiraglia and Lee (2012) demonstrate the manner in which cataloging practice made the designation of authorship something of an obsession.

8.1.4 From Intellectual Content to Resource Description

So we see that a primary function of metadata in information retrieval is to gather, disambiguate, and point to works, which are individual syntheses of recorded knowledge. Works are recorded with text and are found in documents, all of which have certain evidentiary status. The role of metadata is to gather works under their socially constructed historical anchors (epistemologically speaking), which is quite a different thing from making legal attribution. In the role of disseminating culture, the concept-theoretic as applied to works means the creation of alphabetico-classed structures for authors and their works. We turn next to the description of physical resources. A recent textbook by Miller (2011) contains practical summaries and instructions for the use of all forms of metadata for resource description.

8.2 Metadata for Resource Description

Here it is our objective simply to point to the research that has sought to provide crosswalks—or filters—for the large variety of metadata schema that have arisen in the digital age. Howarth (2000), Howarth et al. (2002), Howarth (2003), and Howarth and Miller (2006) are the seminal papers. Prior to this work, research in metadata largely was concerned with development. Howarth (2000) reported some of the first research that could be said to have engaged metadata at a meta-analytical level. That is, she attempted to "crosswalk" across metadata schemas to demonstrate what would be required to move from machine-understandable iterations to human-understandable. Her intention was to develop a meta-level ontology that could be used as a switching device for searchers. She discovered little terminological congruence across schemas, thus demonstrating both the need for such a switching device and the challenge its development would represent. Interestingly, she reported that 68 % of the metadata elements in seven schemas were unique—that is, did not overlap other schemas, further pointing to the evolving complexity of metadata development for resource description. Subsequently Howarth et al. (2002) reported on the attempt to create the proposed switching device. Employing a technique of creating common namespaces for relating semantic commonalities among now nine metadata schema, eighteen categories were devised. These were categories such as (2002, 229): Contact information, Date & time period, Genre type, Language, Methodology, Rights & restrictions on use, Sources, references & related works, and Title, for example.

The crosswalk developed by this team became a primary resource for research. Where many earlier projects contributed substantial data about resource description and metadata generation, none had attempted to provide empirical evidence derived from information objects to underpin their developments. In fact, she and her successive research teams developed a very valuable methodology termed "namespace compilation." This technique yielded a mediating set of common category labels that were subsequently tested using focus groups of end-users. The next step was the development of a language-independent search prototype, for cross-language information retrieval. Using a test repository of documents concerning British Women Romantic Poets from the University of California at Davis, the team developed a technique for visual representation of query results that uses the common categories visually to guide users through the results of their queries, which can be independent of language. Here we see evidence of high-impact, grant-funded research, of the leadership of successive research teams, of a research agenda successfully pursued and evolving with care over time. Here we see also the much needed contribution of empirical evidence to the development and use of metadata schema, as well as very helpful indicators of the direction required for cross-repository and cross-domain use of information resources. Howarth (2006) identifies a gap in the literature—to wit the historical interweaving of evolving trends in metadata and bibliographic control—and very helpfully lays out sets of parallel and interweaving developments that have led toward today's search for increasing interoperability.

8.3   Metadata of Other kinds

Finally, every time someone creates a folder name to save a file on their computer, every time someone writes a filename on the tab of a file folder, every time someone titles a document—all of these are examples of metadata for resource description. If knowledge organization expects as a discipline ever to constitute the concept-theoretic for resource description, then much research of a social nature will be required. For the moment we will point simply to Ferraioli (2005), in which temporal, spatial and contextual factors influence in the creation of personal metadata are explored.

References

Barthes, Roland. 1975. *The pleasure of the text*; trans. by Richard Miller with a note on the text by Richard Howard. New York: Noonday Press.
Buckland, Michael K. 1997. What is a document? *Journal of the American Society for Information Science* 48: 804–9.
Ferraioli, Leatrice. 2005. An exploratory study of metadata creation in a health care agency. *Cataloging & classification quarterly* 40n3/4: 75–102.

Foucault, MIchel. 1984. What is an author? In *Foucault reader* ed. P. Rabinow. Harmondsworth: Penguin, pp. 101–20.

Greenberg, Jane. 2005. Understanding metadata and metadata schemes. *Cataloging and classification quarterly* 40n3/4: 17–36.

Hjørland. Birger. 1997. *Information seeking and subject representation: an activity-theoretical approach to information science.* New directions in information management 34. Westport, Conn.: Greenwood Press.

Hjørland, Birger. 2003. Fundamentals of knowledge organization. *Knowledge organization* 30: 87–111.

Howarth, Lynne. 2000. Designing a "human understandable" metalevel ontology for enhancing resource discovery in knowledge bases. In Beghtol, Clare, Lynne C. Howarth and Nancy J. Williamson, eds., *Dynamism and stability in knowledge organization, Proceedings of the Sixth International ISKO Conference 10–13 July 2000 Toronto, Canada.* Würzburg: ERGON Verlag, pp. 391–98.

Howarth, Lynne C. 2003. Designing a common namespace for searching metadata-enabled knowledge repositories: an international perspective. *Cataloging & classification quarterly* 37n1/2: 173–85.

Howarth, Lynne C. 2005. Metadata and bibliographic control: soul-mates or two solitudes? *Cataloging & classification quarterly* 40n3/4: 37–56.

Howarth, Lynne C., Hannaford, Julie, and Cronin, Christopher. 2002. Designing a metadata-enabled namespace for accessing resources across domains. In Howarth, Lynne C., Christopher Cronin, and Anna T. Slawek, eds. *Advancing knowledge, expanding horizons for information science: proceedings of the 30th Annual Conference of the Canadian Association for Information Science 30 May-01 June 2002, Faculty of Information Studies, University of Toronto.* Toronto: Canadian Association for Information Science, pp. 223–32.

Howarth, Lynne C. and Miller, Thea. 2006. Visualizing search results from metadata-enabled repositories in cultural domains. In Maicher, Lutz and Park, Jack eds. *Charting the topic maps research and applications landscape: First International Workshop on Topic Maps Research and Applications, TMRA 2005, Leipzig, Germany, October 2005.* Heidelberg: Springer-Verlag, pp. 263–70.

International Federation of Library Associations. 1998. *Functional requirements for bibliographic records.* UBCIM publications–new series 19. München: K.G. Saur. http://www.ifla.org/VII/s13/frbr/frbr.htm or http://www.ifla.org/VII/s13/frbr/frbr.pdf

Lubetzky, Seymour. 1969. *Principles of cataloging: final report. Phase IL Descriptive Cataloging.* Los Angeles: Institute of Library Research, pp. 11–15. [Reprinted in Carpenter, Michael and Elaine Svenonius eds. 1985. *Foundations of cataloging: a sourcebook.* Littleton, Colo.: Libraries Unlimited, pp. 189–91.]

Miller, Steven J. 2011. *Metadata for digital collections. : a how to do it manual.* New York: Neal-Schuman.

Panizzi, Antonio. [1848] 1985. Mr. Panizzi to the Right Hon. the Earl of Ellesmere.—British Museum, January 29, 1848. Reprinted from *Appendix to the report of the commissioner appointed to inquire into the constitution and management of the British Museum.* In Carpenter, Michael and Svenonius, Elaine eds., *Foundations of descriptive cataloging.* Littleton, Colo.: Libraries Unlimited, pp. 18–47.

Smiraglia, Richard P. 2001a. *The nature of a 'work': implications for the organization of knowledge.* Lanham, Md.: Scarecrow Press.

Smiraglia, Richard P. 2001b. Works as signs, symbols, and canons: the epistemology of the work. *Knowledge organization* 28: 192–202.

Smiraglia, Richard P. 2006. Empiricism as the basis for metadata categorization: expanding the case for instantiation with archival documents. In Budin, Gerhard, Christian Swertz and Konstantin Mitgutsch, eds. *Knowledge organization and the global learning society; proceedings of the 9th ISKO International Conference, Vienna, July 4–7 2006.* Würzburg: Ergon Verlag, pp. 383–88.

Smiraglia, Richard P. 2007a. The "works" phenomenon and best selling books. *Cataloging and classification quarterly* 44n3/4: 179–95.
Smiraglia, Richard P. 2007b. Bibliographic families and superworks. In Taylor, Arlene G. ed. *Understanding FRBR: what it is and how it will affect our retrieval tools*. Westport, Conn.: Libraries Unlimited, pp. 73–86.
Smiraglia, Richard P. 2008. A meta-analysis of instantiation as a phenomenon of information objects. *Culture del testo e del documento* 9n°25: 5–25.
Smiraglia, Richard P. and Lee, Hur-Li. 2012. Rethinking the authorship principle. *Library Trends* 61 no. 1: 35–48.
Smiraglia, Richard P., Hur-Li Lee and Hope A. Olson. 2010. The flimsy fabric of authorship. In V. Nesset, Ménard, E. and Mas, S., eds., *Information science: Synergy through diversity: Proceedings of the 38th Annual CAIS/ACSI conference, June 2–4, 2010, Concordia University, Montreal, Quebec, Canada*. Toronto: CAIS. Available http://www.cais-acsi.ca/conf_proceedings_2010.htm.
Smiraglia, Richard P., Hur-Li Lee and Hope A. Olson. 2011. Epistemic presumptions of authorship. In *iConference 2011, inspiration, integrity, and intrepidity (February 8–11, 2011)*. New York: ACM, pp. 137–43.
Svenonius, Elaine. 2000. *The intellectual foundation of information organization*. Cambridge, Mass.: MIT Press.
Wilson, Patrick. [1987] 1989. The second objective. In *The conceptual foundations of descriptive cataloging*, ed. by Elaine Svenonius, 5-16. San Diego: Academic Press.

Chapter 9
Thesauri

9.1 KOS in Natural Language

A thesaurus is a complete knowledge organization system structured in natural language instead of according to its ontological construct. That is, the elements in a thesaurus are given in alphabetical order. Each term is then accompanied by a set of relational indicators that show its place in the overall hierarchy. Thesauri can be faceted, when terms from several facets are chosen and entered into a system in a string. Thesauri increasingly are multi-lingual to accommodate complex cultural demands. In this very brief chapter we will look at the basics of thesaurus construction.

One way to think about a thesaurus is to think of a classification turned inside out. Here is a simple classification for cats:

1.0 Cats
1.1 Domestic long-hair
1.2 Manx
1.3 Persian
1.4 Siamese
2.0 Dogs

You will notice that there is a hierarchy for kinds of cats. You will notice that there is a related hierarchy for dogs, which defines the boundary of the classification for cats—a dog cannot be a cat and vice versa. And you will notice that this classification, although limited, is expansible—we could add other kinds of cats and their species as literary warrant demands if our collection grows.

You also can see that the hierarchy is explicit in the symbolic notation. In a thesaurus we would have this arrangement:

Cats

RT Dogs
NT Domestic long-hair

© Springer International Publishing Switzerland 2014
R.P. Smiraglia, *The Elements of Knowledge Organization*,
DOI 10.1007/978-3-319-09357-4_9

NT Manx
NT Persian
NT Siamese

Dogs

RT Cats

Domestic long-hair (cats)

UF Long-hair (cats)
BT Cats
RT Manx (cats)
RT Persian (cats)
RT Siamese (cats)
Long-hair (cats) USE **Domestic long-hair (cats)**

Manx (cats)

BT Cats
RT Domestic long-hair (cats)
RT Persian (cats)
RT Siamese (cats)

Persian (cats)

BT Cats
RT Domestic long-hair (cats)
RT Manx (cats)
RT Siamese (cats)

Siamese (cats)

BT Cats
RT Domestic long-hair (cats)
RT Manx (cats)
RT Persian (cats)

As you can see, each individual term becomes a searchable node in a thesaurus. Where the classification made the context "cats" explicit by its hierarchy, qualifiers have had to be used in the thesaurus where single terms might otherwise be confused. And each term gives instructions about how to move up, down, or across a hierarchy. Two more kinds of instruction can be found in thesauri: UF or "Used For" and its opposite number "Use," and SA for See Also. "Used For" directs a user away from a term that is not used to the term that has been used. See Also is used occasionally when some user might be interested in an otherwise unrelated term. "Scope" notes are used to describe the conceptual content of a term.

9.2 Thesaurus Construction

Construction of a thesaurus is the deliberate construction of a controlled vocabulary. The vocabulary will be used to store data. For example, it might be used as a knowledge management tool in an office. Used in this way it would provide folder headings for workers' desktop document collections, and it would also serve as a rough classification tool for workers to use in keeping track of each other. A thesaurus can also be used for information retrieval—for assigning descriptors to documents, and for querying a system for their retrieval. The final structure of any thesaurus must be dependent on these primary functional considerations. One useful starting point is the concept of specificity. The level of language to be employed is related to the intended functionality of the thesaurus.

Specificity is related to the intended audience. Formal language might be used for a scientific purpose (such as a medical form), but a frequently chosen alternative is operational language. This simply means the language needed to function in any given work environment. Sometimes thesauri conceived at a meta-level (that is, a very broad overview) turns out to require major parts to be segmented into their individual parts—in this case we say the language is componential. Coercive language might be employed when the purpose is to introduce change into an environment. And, clearly, a useful alternative to all of the above is to generate consensus language. We consider these decisions related to specificity, because the more formal the language the more specific and precise the terms must be, whereas the more informal and consensus oriented the language, the less specific and precise terms will be. Incidentally, none of these concepts is mutually exclusive—several may be used at once within a given thesaurus, if segments so demand.

The designer of a thesaurus must make decisions about how to regularize terminology in the vocabulary (hence the name, controlled vocabulary). Common issues are:

Single words vs. Multi-word terms—consider the difference between the words 'fire' and 'engine' and the term 'fire engine'
Singular (processes and properties) vs. plural (classes of things)—consider Architecture (a process) and Architectural drawings (a class of things)
Direct vs indirect entry—enter under the noun or its modifier? decide based on desirable collocation—Spots, Hot or Hot spots; Music, Popular or Popular Music

Semantics concern meaning, especially meanings among terms in the vocabulary at large. Meaning may be denotative or connotative, for example. Formal vocabularies generate more denotative terms; informal generate more connotative. "House" denotes a dwelling place, but it connotes security. Terms must be distinguished thus in a thesaurus—"House" for a dwelling, "Home" for a private residence, for instance. Semantics must also be considered in the presence of synonyms and homonyms. Synonyms will be distinguished with Use for references; homonyms will require

qualifiers—Bridge (dental). Hierarchies will be ubiquitous in a thesaurus, displaying genus/species and whole/part relationships. Affinity affects semantic relationships among terms that have similar meanings: consider 'teach' and 'teacher' for
example. The relationship is not hierarchical, therefore both terms might find a
home in the same list.

Syntax will be most important in the decision to pre- or post-coordinate terms.
Again, consider 'fire engine,' alongside 'fire house,' and 'fireplace.' Pre-coordination
is more user friendly; post-coordination is more economical.

Is your term list entirely hierarchical? Probably not. Often, thesauri are divided
into facets. A facet is a component part. Usually facets are mutually exclusive, but
collectively exhaustive. That means, no term fits in more than one facet, but all of
the facets together completely describe the domain. A common set of facets is functional: Actors, Actions, Places, Times, Objects acted upon, etc. This is a simple list,
but you might find it helpful. Take your term list and set aside every term having to
do with places, then with times. Now set aside all actions. What do you have left?
Probably actors (people, or governments, or some sort of group) and specific objects.
Each can be called a facet; each facet can probably be arranged hierarchically. What
you will need is a rule of how to use them. Sometimes an order is specified.
Sometimes one is simply encouraged to select terms from each facet.

Properties—my flag is red white and blue (colors) and made of cloth (material).
Those are properties of the entity 'flag.' You might want to divide any facet into entities and properties. How about actors—people (male or female are gender properties, English or Spanish-speaking are language properties)?

9.3   Thesaurus Construction as a Domain

There is quite a large literature on the practicalities of thesaurus construction. The
most complete analysis of relevant research is in Roe and Thomas (2004). The most
concise manual for thesaurus construction is Aitchison et al. (2004). Dextre Clarke
(2001) defined the essential relationships in thesauri, and Zeng and Chan (2003)
surveyed interoperability. The major challenge of multi-lingual thesauri, making
controlled vocabulary work across language boundaries, has been the province of
authors from multi-lingual cultures (see Hudon 1997).

References

Aitchison, Jean, Alan Gilchrist, and David Bawden. 2004. *Thesaurus construction and use:
     A practical manual.* 4th ed. London and New York: Europa.
Dexter Clarke, Stella. 2001. Thesaural relationships. In Bean, Carol, and Rebecca Green, eds.
     Relationships in the organization of knowledge. Netherlands: Kluwer, pp. 37–52.
Hudon, Michele. 1997. Multilingual thesaurus construction—integrating the views of different cul-
     tures in one gateway to knowledge and concepts. *Information services and use* 17: 111–23.

Roe, Sandra K., and Alan R. Thomas, eds. 2004. *The thesaurus: review, renaissance and revision.* Binghamton: Haworth. (Also issued as *Cataloging and classification quarterly* 37n3–4).

Zeng, Marcia Lei, and Lois Mae Chan. 2003. Trends and issues in establishing interoperability among knowledge organization systems. *Journal of the American Society for Information Science and Technology* 55: 377–95.

Chapter 10
Domain Analysis

10.1 About Domains

Knowledge organization, as we have seen, is the attempt to ascertain both the natural orderings of knowledge in every domain and the attempt to ascertain the most useful orderings of recorded knowledge for retrieval. We have seen that the concept—a named and defined idea—is the atomist entity for knowledge organization. That is, both the science of knowledge organization and the activity of developing systems for knowledge organization rely on the concept at the core. Therefore the most critical aspect of the field is determining those concepts that are to be ordered in knowledge systems. Domain analysis is the activity, or the methodology, by which the conceptual content and natural or heuristic orderings can be discovered and mapped in discrete knowledge domains. Increasingly, domain-analytic studies are being used to compare domains as well as to track their evolution. We will review these tools and some examples from recent research to see just how domain analysis can be successfully employed in the design of KOS.

There has been some confusion about the definition of a domain, perhaps because the notion must embrace diversity. Some have written of discourse communities, disciplines, invisible colleges, and even work ecologies, and all of these are kinds of domains. Basically a domain is simply a group with a coherent ontology. At a more academic level a domain is a group of scholars working on research problems that are in some way perceived to be similar. The notion of discourse suggests that some sort of social networking takes place among participant scholars. In fact all three points of view combine in the definition of domain. A domain is a group that shares an ontology, undertakes common research or work, and also engages in discourse or communication, formally or informally.

© Springer International Publishing Switzerland 2014
R.P. Smiraglia, *The Elements of Knowledge Organization*,
DOI 10.1007/978-3-319-09357-4_10

Recently Smiraglia (2012) recounted the evolution of the notion of domain in information and knowledge organization, and posed a formal definition. He wrote (p. 114):

A domain is best understood as a unit of analysis for the construction of a KOS. That is, a domain is a group with an ontological base that reveals an underlying teleology, a set of common hypotheses, epistemological consensus on methodological approaches, and social semantics. If, after the conduct of systematic analysis, no consensus on these points emerges, then neither intension nor extension can be defined, and the group thus does not constitute a domain.

We can unpack this definition based on what we have learned in earlier chapters to reveal its operational capacity. We understand first that methodologically a domain is any group that is useful for the construction of a knowledge organization system. That leads naturally to the conditions that the group share knowledge (an ontological base), goals (an underlying teleology), research methods (hypotheses and methodologies), and a functional system for communication (social semantics). This unpacking gives us a lot of leeway. A domain, then, can be a group of people who work together if they share knowledge, goals, methods of operation, and communication. It can be a community of hobbyists, a scholarly discipline, an academic department, and so on. The tools of knowledge organization applied to the products of a domain should reveal these contours in measurable form. This is what we call intension and extension, which are the depth and breadth, respectively, of the shared knowledge base (Tennis 2003).

10.2   About Domain Analysis

The leading advocate of domain analysis is Hjørland, who has called for the use of domain analysis as a core methodology for information, as well for knowledge organization, and who also has elaborated 11 steps that can provide information about a domain. These steps are (2002, 450–51):

- Literature guides—idiographic descriptions of information resources in a domain;
- Special classifications and thesauri— demonstrate logical conceptual structures and semantic relations;
- Indexing—reveals the epistemological potentials of individual documents;
- Empirical user studies—reveal mental models of domain actors;
- Bibliometric studies—demonstrate sociological patterns of explicit recognition among documents;
- Historical studies—to reveal traditions;
- Document and genre studies;
- Epistemological and critical studies;
- Terminological studies;
- Studies of structures and institutions in scientific communication; and,
- Domain analysis in professional cognition and artificial intelligence—to elicit mental models.

In this list we see tactics for discerning the roles of activity and actors, of teleo-logical imperatives, of common ontology, and of the social semantics of any group engaged in intellectual collaboration. Smiraglia (2012) contains a meta-analysis of published domain analytical studies in knowledge organization. In fact, most domain analysis is informetric, using combinations of citation analysis, author co-citation analysis, co-word analysis, and network analysis to compare visualizations of a domain.

10.3 Techniques for Domain Analysis

Methodologically, domain analysis requires mixed methods approaches. Much of the work falls into what is thought of as qualitative by nature, although quite often quantitative techniques also are employed. Perusing Hjørland's list above makes it clear that all of those 11 approaches must take place within a specified environment, whether that be an office with workers or a domain with journals and conference proceedings. Establishing the domain by designating the boundaries for analysis is essentially a subjective task. Selecting parameters means using subjective decision-making to establish boundaries. But once that has been achieved, a number of quantitative measures can be applied. The effective use of ethnographic methods has been demonstrated by Hartel (2003, 2010) who used forms of participant observation to analyze hobbies and serious leisure activity, most notably cooking.

10.3.1 Citation Analysis

The most common informetric methods employed in domain analysis are citation analyses. Citations are a form of social networking. A scholar cites a published paper to designate authority for quoted work, but also as a means of associating research. If I cite a published paper it is a means of associating my work with that of the authors of the other paper. So in a sense it is a form of academic social networking. For that reason citations leave behind trace evidence of associations that might be tightly-woven within the domain, or might reach outside the domain, or both. A thorough source for the bases of citation analysis is De Bellis (2009).

Citations are ubiquitous in scholarly discourse, of course, making them an attractive data source for analysis. But indexing is a critical component of citation research. Large swaths of the sciences, social sciences, and humanities are indexed by two global giants Thomson-Reuters *Web of Science* and Elsevier *Scopus*. If the domain under study is indexed, compiling simple citation statistics can be almost automatic. For example, by searching the *Web of Science* (*WoS*) for "multi-lingual thesauri" a small exemplar of a domain is located including six publications. These are shown in Fig. 10.1.

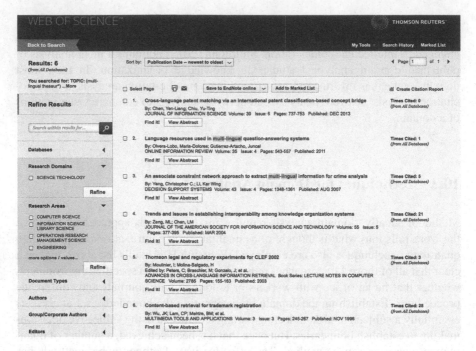

Fig. 10.1 "Multi-lingual Thesauri" from *Web of Science*

Automatically, given specific parameters, *Web of Science* produces a list of all publications in its database that meet the criteria. A variety of analyses are available at once, including categories and document types that are shown in the pull-down menus at the left; together with publication years and document types these are shown in Fig. 10.2.

By clicking on "Create Citation Report" at the upper right on the first result screen one can be taken to a set of data based on the citation analysis of this topical domain. Some simple metrics are given in tabular form for dates of works in the result set and citations to the source documents in the result set shown over time (Fig. 10.3).

On the same screen *Web of Science* displays citation data about the six source items; these include the total number of citations for the group as well as for each source document and the average number of citations per year, as shown in Fig. 10.4.

By clicking on the blue numerals in the "Total" column, one can be taken to a display of the works that cite that source for further detailed analysis. This is deliberately a very simple example to demonstrate the utility of beginning domain analysis within one or the other of the major citation indexing services. As we have seen it was possible to define the domain's geographic and chronological demographics, the citing patterns of the authors, and the chronology and impact of citations to the source documents.

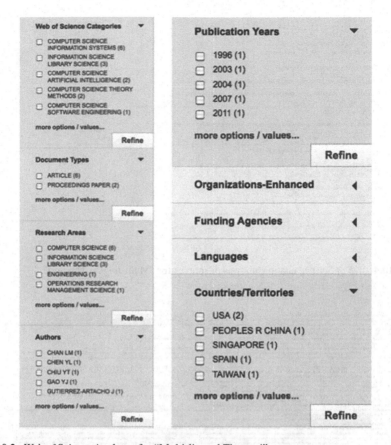

Fig. 10.2 *Web of Science* Analyses for "Multi-lingual Thesauri"

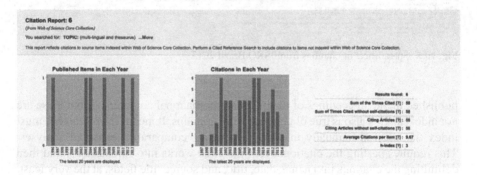

Fig. 10.3 *Web of Science* Citation Report for "Multi-lingual Thesauri"

However, quite a lot of scholarship is not indexed by these services, including much of the domain of knowledge organization. Research has shown (Smiraglia 2014a) that 60 % of the productivity in the knowledge organization domain is

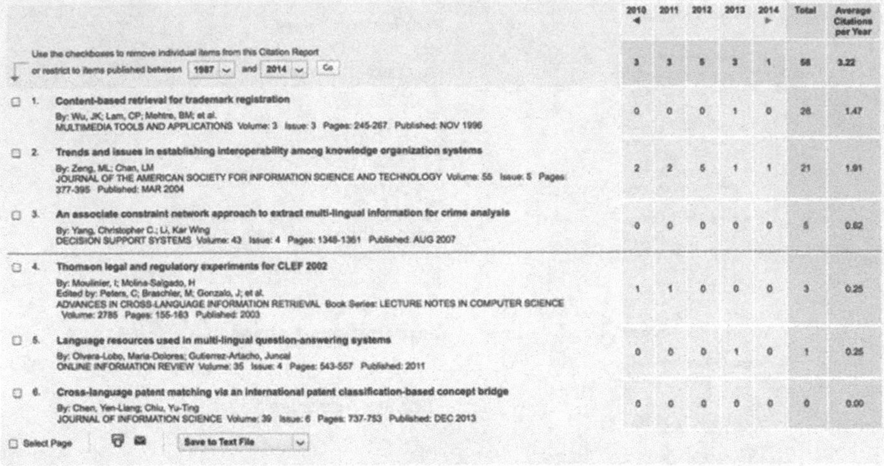

Fig. 10.4 *Web of Science* citation analysis of sources for "Multi-lingual Thesauri"

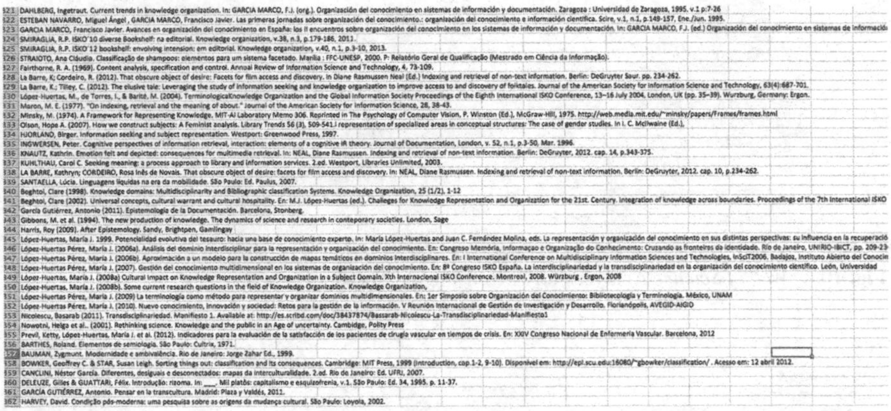

Fig. 10.5 Spreadsheet of citations from ISKO Brazil 2013

published in the proceedings of the biennial international conferences, but these are not indexed. This also is true of new, evolving domains. It means the researcher must index the material manually in order to generate comparable statistical analyses. This requires pasting the citations from the cited works into spreadsheets and then delimiting the citations into name, date, title, and source title fields, at the very least. This often is problematic when the editors of proceedings have not required authors to use a single citation format. Figure 10.5 is a display of a spreadsheet developed for analysis of the 2013 ISKO Brazil conference reported in Smiraglia (2014b).

Some dates are in parentheses in author-date position following the author names, some appear at the end of the citation in a bibliographical style. Some author names are given in full, some are abbreviated. These disparities in publication are a

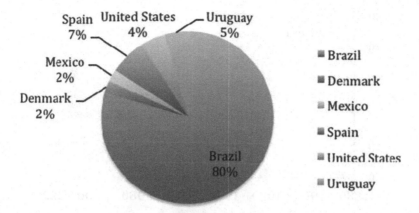

Fig. 10.6 Distribution of geographic affiliation of authors in ISKO Brazil 2013 (Smiraglia 2014b, 106)

serious drawback to research in knowledge organization. Nonetheless, manual analysis can produce a visualization of a domain.

For example, analysis of the papers contributed to ISKO Brazil 2013 showed a mean of 11 references per paper, with a range from zero to 48. Small numbers of citations align with the hard sciences, where research is rapidly cumulative, and large numbers of references align with the humanities, where research is dependent on traditional sources. Eleven references per paper is on the shorter end of the distribution, which is consistent with knowledge organization's alignment with the social sciences such as information. But the range up to 48 shows us that ongoing reliance within the domain on humanistic approaches to theory-driven research. The same analysis divided the citations by country of author affiliation. The geographic orientation is reprinted here in Fig. 10.6.

This shows the geographic influence of the authors taking part in the conference. If the domain is coherent intellectually, this is a remarkable commentary on the global nature of the science. As a matter of fact, there is intellectual coherence, but as we can see, at this Brazilian conference most of the papers come from local authors, which is no surprise.

Often several different measures are compared to one another, as a form of methodological triangulation, which is regarded as a qualitative technique. For example, the dates of publication of works cited by the authors of the conference papers are visualized in the table reproduced in Fig. 10.7.

Now we see a long tail stretching from about 1902 to the present, but we also see that the majority of works cited fall after about 1980. That means the majority of works cited are about 30 years old, which, like the data about numbers of citations, aligns with social scientific domains. But we also see the humanistic influence in the continued citation, if infrequent, of works more than a 100 years old. This is what is meant by methodological triangulation—two different measures both give the same impression of the domain.

The final simple citation measure was the distribution of publication media. This is represented in the table reproduced in Table 10.1.

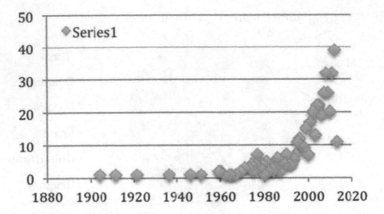

Fig. 10.7 Distribution of dates of publication of works cited in ISKO Brazil 2013 (Smiraglia 2014b, 107)

Table 10.1 Distribution of media types cited in ISKO Brazil 2013 (Smiraglia 2014b, 107)

Monograph	210
Journal article	155
Conference proceedings	31
Other	24
Dissertation (PhD)	18
Website 1	3
Thesis	8
Report	3
Database	2

Here we see that most of the citations are to monographs, which is a humanistic tendency. But, on the other hand, if we add together journal articles and conference proceedings together with dissertations and theses, we have almost an equal number (212) representing likely empirical research. So that shows us, again, that there are both empirical and humanistic approaches at work in this domain. This is the third methodologically triangulated result, which gives us a fair amount of confidence (not statistical confidence mind you) to say that the domain shares both scientific and humanistic epistemological stances, expressed through their methodological approaches to research. This is consistent with other analyses of knowledge organization communities. But it also tells us that this domain is made up of two distinct epistemological communities, sharing goals and a common knowledge base.

10.3.2   Co-word Analysis

Continuing with our example of the ISKO Brazil 2013 conference we can look at a simple demonstration of co-word analysis. Co-word analysis uses software to calculate word or term frequencies in a body of text. For these examples Provalis

Table 10.2 Frequency distribution of title and cited-title keywords in ISKO Brazil 2013 (Smiraglia 2014b, 108)

Conference paper title keyword	Frequency (%)	Cited paper title keyword	Frequency (%)
CONHECIMENTO	21.50	KNOWLEDGE	4.60
ORGANIZAÇÃO	16.50	INFORMATION	4.30
REPRESENTAÇÃO	13.90	INDEXING	3.60
ANÁLISE	6.30	INFORMAÇÃO	3.60
CIENTÍFICA	6.30	ORGANIZATION	3.60
PRODUÇÃO	5.10	SUBJECT	2.90
INFORMAÇÃO	5.10	ANALYSIS	2.60
PESQUISA	5.10	ANÁLISE	2.10
ARTIGOS	3.80	THEORY	2.10
INDEXAÇÃO	3.80	CIÊNCIA	1.90
REFLEXÕES	3.80	CONOCIMIENTO	1.90
SOCIAL	3.80	BRASIL	1.70
INDEXAÇÃO	1.70		
SCIENCE	1.70		
CIENTÍFICO	1.60		
CLASSIFICATION	1.60		
INFORMACIÓN	1.60		
REPRESENTAÇÃO	1.60		
RETRIEVAL	1.60		
CONHECIMENTO	1.40		
DIGITAL	1.40		
TEÓRICO	1.40		

Research's SimStat-WordStat™ software was used to generate analyses. In Table 10.2 we see a distribution of keywords from the titles of the conference papers alongside keywords from the papers cited by the authors of the conference papers.

This is a simple form of data triangulation, in which we compare keywords from the conference papers—assigned by the authors—and keywords from the papers cited by them—assigned by the research domain at large. You see words in Spanish, Portuguese and English, which is characteristic of ISKO Brazil. Also we can combine words into terms, such as "conhecimento organização," Portuguese for knowledge organization. In that manner we can see that the two lists are pretty much the same, indicating a core ontology exists in this domain, reflected both by the authors in the domain and by the papers they cite. More sensitive analysis can be conducted on larger data sets by generating term dictionaries that can be used to filter text, using titles, abstracts, or even full texts of papers.

Co-word analysis also can be used to create a visualization of the core ontology by mapping terms according to co-occurrence statistics. WordStat™ can produce two- or three-dimensional maps using multi-dimensional scaling (MDS) to plot the relative position of terms. A simple display, based on the words in Table 10.2, is shown in Fig. 10.8.

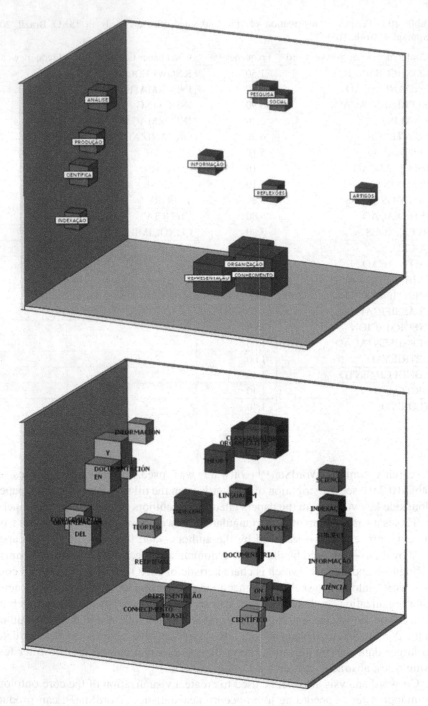

Fig. 10.8   MDS plot of cited-title keywords in ISKO Brazil 2013 (Smiraglia 2014b, 109)

The advantage of visualization comes from providing a graphic overview of the domain. For instance in this case the different colors align with the different language blocks within the keyword list. So we can actually see the comparative impact of the different language contributions to the proceedings under analysis.

10.3.3   Author Co-citation Analysis

Author co-citation analysis is based on the idea that if two authors are citing the same material they likely are engaged in similar or comparable research, or at the very least are working in the same domain. Using techniques clearly explained by McCain (1990), author co-citations among the works cited by a domain can be used to generate visualizations of co-cited authors. These visualizations represent the view of the domain held by those who co-cite these authors. The technique involves gathering co-citation data into a matrix and then processing it using software that can create MDS plots based on various co-occurrence statistics. Common are IBM-PAWS SPSS™ MDS plots like those shown in Fig. 10.9.

Here we have yet another form of triangulation. First, author-cocitation plots generate thematic clusters. The hope is that these thematic clusters when analyzed will be similar to those visualized using co-word analysis. Where there is divergence between the two methods, additional information about the domain can be derived. The second form of triangulation occurs when we look at both how the domain is viewed by authors whose papers are in the *Web of Science* (the upper plot in Fig. 10.9), versus how the domain is viewed by the authors contributing papers to the conference (the lower plot in Fig. 10.9). Obviously the second plot requires manual processing of the citations to find author cocitation. Here is how these plots were analyzed in publication (Smiraglia 2014b, 110–11):

> [In the upper plot] the cluster on the right clearly represents traditional knowledge organization, and the cluster on the left mixes influences from Spanish, Brazilian and American authors .... [The lower plot] is rather a different map. Here we see more or less the same groupings but differently positioned, and with an epistemological axis implicitly ranging from the empirical on the left to the rational and historicist on the right .... There is a clear integration of traditional knowledge organization concepts. But there also is original work from the Americas, particularly with regard to indexing, documentation and informetrics.

10.3.4   Network Analysis

Network analysis can lead to more complex visualizations that offer yet another view of a domain. Network theory is a way of mapping relationships among objects in a data set based on the symmetry or asymmetry of their relative proximity. The network map in Fig. 10.10 represents a visualization of network relationships among the co-cited authors in the Brazil conference papers (the lower map from Fig. 10.9).

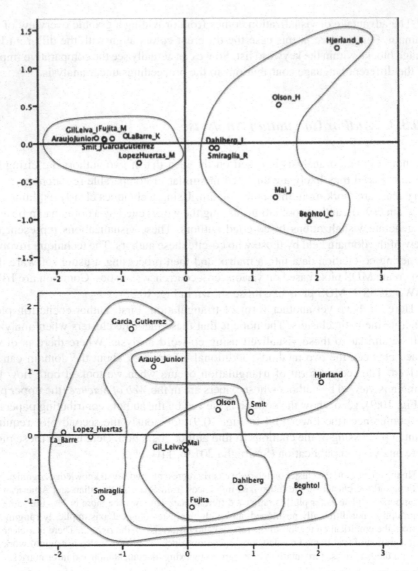

Fig. 10.9 MDS plot of author-cocitation from ISKO Brazil 2013 (Smiraglia 2014b, 110–11). The upper plot shows author-cocitation from the *Web of Science*; the lower plot shows author-cocitation within the proceedings of the conference

The complexity of the network map helps us visualize the degree of interconnectedness among the thematic clusters represented by co-cited authors. The different densities of the connecting edges helps us visualized the relative strength of the associations.

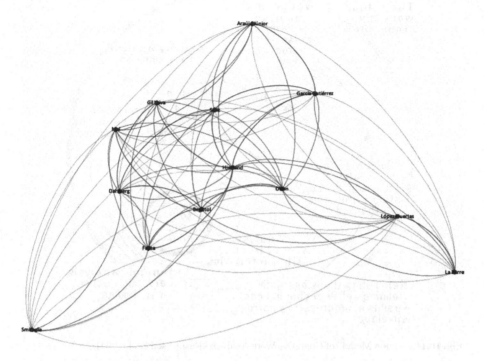

Fig. 10.10 *Gephi*™ network diagram of interconference author cocitation from ISKO Brazil 2013 (Smiraglia 2014b, 111)

10.3.5 Cognitive Work Analysis

Cognitive Work Analysis (CWA) is a relatively new method for domain analysis brought most effectively to the knowledge organization domain from work by Rasmussen et al. (1994), who generated the famous "onion model" shown in Fig. 10.11.

The methodology is essentially qualitative borrowing from ethnographic research. The researcher goes inside a work environment (the domain) to learn from all of the participants (actors) how they interact, how they interact with clients outside the domain, how they generate and share knowledge, and how they organize their work. Means-ends analysis is applied to the data to generate results visualizing the shared ontology as well as its task-based heuristics. An early study in knowledge organization to use CWA was Albrechtsen et al. (2002), in which collaborative film-indexing was the domain under study. A concise report by Albrechtsen and Pejtersen (2003) appeared in the journal *Knowledge Organization* to explain the use of the technique for generating work-based classifications. Mai (2008) extended the model to suggest means of using it to generate controlled vocabularies.

A recent study by Marchese (2012) used CWA to analyze the knowledge base of a Long Island HR firm. This report is remarkable for its clear explanation of the methodology. Figure 10.12 is an illustration of the work environment made by the researcher while inside the domain.

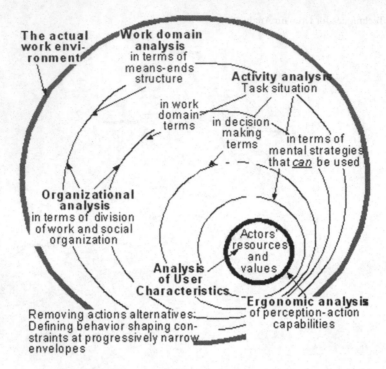

Fig. 10.11 "Onion Model" of Cognitive Work Analysis (Rasmussen et al. 1994)

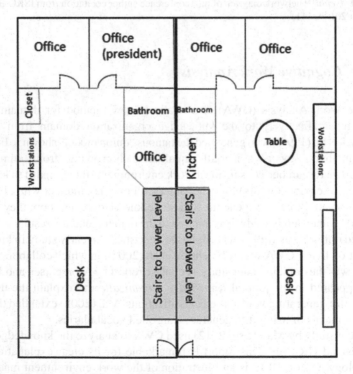

Fig. 10.12 Office floor-plan (Marchese 2012, 63)

The floor plan became an important part of the visualization of the knowledge base because it turned out there were different vocabularies shared in the open spaces than in the closed offices, and also there was a different vocabulary used for communicating outside the staff with clients. The emergent vocabulary including "boundary objects," or terms used to pivot from one vocabulary to another, say from inside to outside, were explored in Marchese and Smiraglia (2013). This is reproduced in Table 10.3.

Table 10.3 CWA revealed emergent vocabulary showing boundary objects (Marchese and Smiraglia 2013, 255)

Articulate	Effective	Pipeline
Break-out groups	Efficient	Process
Broader audience	Employee levels	Report out
Buckets	Executive development, Learning development,	Results
Business skills	Focus groups	Roll-up of data
Characters/role play	Gap scores	Rotate
Check-ins	Individual behavior	Share methodology
Cleaner	Interviews	Step-back
Client's chart preference	Learning styles	Strong
Data	Logs	Super days
Descriptive	My lead – meetings -> product , task	Surveys
Developmental priorities	Organizational Behavior	Team behavior
Diversity	Phone bank	Thought process
Divisions, levels, products, job families, business units		

The highlighted terms are those that are shared between insiders and outsiders, and thus represent boundary objects, or points of opportunity for creating interoperable neighboring vocabularies from shared ontologies.

10.4   The Role of Domain Analysis

As we have seen, domain analysis can produce a wealth of information about the ontological functioning of a community. In particular it can be used to generate knowledge organization systems, such as controlled vocabularies or classifications, to assist the domain in its work. Perhaps more important to our post-modern world, domain analytical studies can produce the evidence needed to provide interoperability between neighboring domains and among diverse domains. In addition to domain analytical work for knowledge organization systems, the same methods have been used to track the evolution of domains across time (Smiraglia 2009a, b). This research can provide background for what has been called subject ontogeny (Tennis 2002, 2007), in which the relative positions of ontological concepts can be traced back across time to observe semantic evolution. The importance of domain analysis for knowledge organization as a science cannot be overlooked.

References

Albrechtsen, Hanne, Pejtersen, Annelise Mark and Cleal, Bryan. 2002. Empirical work analysis of collaborative film indexing. In Bruce, Harry, Raya Fidel, Peter Ingwersen, and Pertti Vakkari, eds., *Emerging frameworks and methods: Proceedings of the Fourth International Conference on Conceptions of Library and Information Science*. Greenwood Village, CO: Libraries Unlimited, pp. 85–108.
Albrechtsen, Hanne and Pejtersen, Annelise Mark. 2003. Cognitive Work Analysis and work centered design of classification schemes. *Knowledge organization* 30: 213–27.
De Bellis, Nicola. 2009. *Bibliometrics and citation analysis: from the* Science Citation Index *to Cybermetrics*. Lanham, Md.: Scarecrow.
Hartel, Jenna. (2003). The serious leisure frontier in library and information science: hobby domains. *Knowledge organization* 30: 228–38.
Hartel, Jenna. 2010. Managing documents at home for serious leisure: a case study of the hobby of gourmet cooking. *Journal of documentation* 66: 847–74.
Hjørland, Birger. 2002. Domain analysis in information science: eleven approaches – traditional as well as innovative. *Journal of documentation* 58: 422–62.
McCain, Katherine W. 1990. Mapping authors in intellectual space: a technical overview. *Journal of the American Society for Information Science* 41: 433–43.
Mai, Jens-Erik. 2008. Design and construction of controlled vocabularies: analysis of actors, domain, and constraints. *Knowledge organization* 35(1), 16–29.
Marchese, Christine. 2012. Impact of organizational environment on knowledge representation and use: cognitive work analysis of a management consulting firm. Ph.D. dissertation. Long Island University.
Marchese, Christine, and Richard P. Smiraglia. 2013. Boundary objects: CWA, an HR Firm, and emergent vocabulary. *Knowledge organization* 40: 254–59.
Rasmussen, Jens, Annelise Mark Pejtersen and L.P. Goodstein. 1994. *Cognitive systems engineering*. New York, NY: Wiley.
Smiraglia, Richard P. 2009a. Modulation and specialization in North American knowledge organization: visualizing pioneers. In Jacob, Elin K. and Barbara Kwasnik, eds., *Pioneering North American contributions to knowledge organization, Proceedings of the 2d North American Symposium on Knowledge Organization, June 17–18, 2009*, pp. 35–46 http://dlist.sir.arizona.edu/2630/

Smiraglia, Richard P. 2009b. Redefining the 'S' in ISMIR: visualizing the evolution of a domain. In Rothbauer, Paulette, Siobhan Stevenson, and Nadine Wathen, eds. *Mapping the 21st century information landscape: borders, bridges and byways: Proceedings of the 37th Annual CAIS/ACSI Conference, May 28–30, 2009, Ottawa, Ontario, Canada.* http://www.cais-acsi.ca/proceedings/2009/Smiraglia_2009.pdf

Smiraglia, Richard P. 2012. Epistemology of Domain Analysis. In Smiraglia, Richard P. and Hur-Li Lee eds. 2012. *Cultural frames of knowledge.* Würzburg: Ergon Verlag, pp. 111–24.

Smiraglia 2014a. Meta-analysis: the epistemological dimension of knowledge organization. *IRIS Revista de Informação, Memória e Tecnologia.* forthcoming.

Smiraglia 2014b. II Congresso Brasileiro em Representação e Organização do Conhecimento: Knowledge Organization in Rio 2013—An Editorial. *Knowledge organization* 42: 105–12.

Tennis, Joseph T. 2002. Subject ontogeny: subject access through time and the dimensionality of classification. In María José Lopez-Huertas, ed. *Challenges in knowledge representation and organization for the 21st century: integration of knowledge across boundaries: Proceedings of the Seventh International ISKO Conference, Granada, 10–13 July 2002.* Würzburg: Ergon Verlag, pp. 54–59.

Tennis, Joseph T. 2003. Two axes of domains for domain analysis. *Knowledge organization* 30: 191–5.

Tennis, Joseph T. 2007. Scheme versioning in the semantic web. *Cataloging & classification quarterly* 43 no. 3: 85–104.

Printed in the United States
By Bookmasters

Printed in the United States
By Bookmasters